After the Big Bang Could Come the Big Crunch

Primordial Black Holes and the Big Crunch

By Solomon M. Jacob, PhD

DORRANCE
PUBLISHING CO
EST. 1920
PITTSBURGH, PENNSYLVANIA 15238

Dorrance Publishing Co
585 Alpha Drive
Suite 103
Pittsburgh, PA 15238
Visit our website at *www.dorrancebookstore.com*

ISBN: 979-8-88729-139-0
eISBN: 979-8-88729-639-5

After the Big Bang Could Come the Big Crunch

Primordial Black Holes and the Big Crunch

SYNOPSIS

Like the Universe itself, our understanding of the Universe is (hopefully) always expanding!

This book explains what happened at the creation of the universe: the so-called 'Big Bang'; discussing the many events that took place within the first few milliseconds of the life of the universe. Readers are invited to consider how life exists both on a scale that is infinitely large (the planets and stars that make up the universe), down to one that is minutely small (elementary particles); and to consider how the theories behind these concepts can be reconciled.

A novel description of the Universe is presented in this book: it is based on an immediate BIG CRUNCH (BC) that followed the Initial Cosmic BIG BANG (BB). Due to the formation of Primordial Black Holes (PBHs) occurring almost simultaneously to the BB, a HUGE amount of Energy was taken back into the PBHs. We are calling this re-absorption, "the BIG CRUNCH"; while acknowledging, along with the recent Bellini & Garcia* postulation, that the PBHs themselves are the leading candidates for Dark Matter in our Universe (which makes up 28% of the Universe). This model further accurately predicts that because of the BC, we will have 72% dark energy (supported by the existence of at least two concentric rings recently found in the Cosmic Microwave Background*). To conclude, this model is highly consistent with the generally accepted concept that the universe is made up of 28% dark matter and 72% dark energy. Remarkably, this means that

EVERYTHING that we can see, such as our galaxy and other galaxies, make up just approximately 5%, (4.8%), of the Universe!

A clear explanation is given of many hitherto inaccessible concepts such as cosmological inflation, Hawking radiation, primordial black holes and quantum mechanics. The universally accepted ideas about the evolution of the universe, as well as those that are more bizarre and outlandish, are presented.

Ultimately, readers are invited to think about whether and how our universe experienced the "BIG CRUNCH." If proven correct, then this book has the potential to completely change our view of the universe.

Disclaimer: This book will have many detractors and critics because the model we propose is not based on Einstein's theory of General Relativity. The equations of general relativity are highly nonlinear, and are almost impossible to solve, even with the most sophisticated supercomputers. The model we propose is based on Newton's equations of mass and energy- and are easy to solve. We strongly believe that this model is directionally correct and could be nearly 100% correct. Only time will tell how accurately this model explains dark matter, dark energy and the expansion of the Universe!

PROLOGUE

It was a "starry starry night" in January 2011 that the concept for writing this book came into being. I was vacationing on our houseboat on lake Eildon (on the outskirts of Melbourne, Australia) with my wife Rita and my identical twin daughters Majah and Sarah. It was a particularly dark night with no moon, and we were shielded from the light of the megalopolis of Melbourne. As we lounged on the roof of our houseboat & looked up, it was breathtaking!

Even with the naked eye, we could clearly see the galaxy of Andromeda and the Magellanic clouds. Other fainter galaxies could be seen as dabs of matter in the sky. Globular clusters, which are made up from stars, were also visible. For the first time ever, I could feel the rotation of the Earth by staring at the stars in relation to the dead branches of trees surrounding the lake.

It was on that night that I decided to learn a lot more of how the universe evolved. I started reading several books by Stephen Hawking on black holes and particle physics. Many other books were read, like Brian Greene's on the creation of the universe, and another famous book by Alan Guth on the first three minutes since the Big Bang and the concept of inflation, (the event that somehow helped create the universe as we know it today). No one actually knows how this happened and there are many theories out there that all try to explain this phenomenon, some of which are really bizarre and farfetched!

This book attempts to explain the evolution of the universe for the many readers out there that are interested in this subject!

by Solomon Jacob, Ph.D.

Acknowledgments

I want to acknowledge a lot of people, including my grandson Bodhi, for insisting that I write down the Newtonian equations instead of just waving my hands. My grandchild Pai, in the early writing stage of the book, sent me various clips from the internet on how the Universe evolved. My granddaughter Lilah for her help in editing the text. I also thank Julia Avery for many, many suggestions as to how to improve the book, including her expert contributions on editorial and contextual matters. Cristina Pelka for the well-composed illustrations with an aesthetic touch. And finally, I would like to thank my daughters Majah and Sarah for their helpful comments on the text, especially in the introduction.

Table of Contents

INTRODUCTION

CHAPTER 1
LIFE AND THE UNIVERSE AS WE KNOW IT TODAY

1. OUR UNIVERSE

The universe is a big place, and only getting bigger with time. In fact, scientists have long known that the universe is expanding but, only recently, have they found that this expansion is also accelerating. It is not only the biggest thing in existence but also the oldest. Researchers now have an accurate estimate of the age of the universe, 13.7 billion years old, give or take 20 million years.

Our galaxy, the Milky Way, one of trillions in the observable universe was formed approximately 9 billion years ago. The Earth was then formed shortly after some 4.7 billion years ago. Then, life found its way to Earth.

> *"We know that life began at least 3.5 billion years ago, because that is the age of the oldest rocks with fossil evidence of life on earth. These rocks are rare because subsequent geologic processes have reshaped the surface of our planet, often destroying older rocks while making new ones. Nonetheless, 3.5-billion-year-old rocks with fossils can be found in Africa and Australia. They are usually a mix of solidified volcanic lavas and sedimentary cherts. The fossils occur in sedimentary cherts."*
>
> – David A. Kring

Here we are billions of years later trying to figure it all out. Although we have come a long way to understanding aspects of the evolution of the universe, there are still big problems in the way. Some of these problems are so significant that seldom few scientists even find the courage to tackle them directly. There is still a long way to go, indeed!

2. The Speed of Earth

If you are standing on the ground, motionless, then what is your speed? I bet your answer is zero, right! Well, that's not the full story. Your relative speed, with respect to the buildings and stationary objects around you might be zero, but the Earth itself is moving through space at a tremendous speed. In fact, it not only rotates like a spinning-top but also revolves around the Sun. Moreover, our solar system is moving as well. This may seem strange to you because there is no sensation of speed, the type of thing that is felt in a speeding car.

"We're travelling at 1,332,000 kilometers an hour, says Mike Hudson of the Department of Physics and Astronomy at the University of Waterloo. The sun is moving at 1,332,000 kilometers an hour compared with a distant horizon defined by the cosmic microwave background radiation. This speed is made up of three velocities – the sun's motion around the centre of the Milky Way galaxy (792,000 km/h), the Milky Way's speed while approaching our nearest neighbor, Andromeda (442,800 km/h), and the speed at which both the Milky Way and Andromeda galaxies are moving compared with the background radiation (2,160,000 km/h). These three speeds don't add up to 1,332,000 km/h, however, because the velocities are heading in different directions. Some add, some partly add and some partly subtract. Meanwhile, the speed at which the Earth rotates on its axis (1,656 km/h at the equator) and the Earth's orbital velocity around the sun (108,000 km/h) are relatively small." – Philip Jackman (2012)

It is important to understand that these speeds, although staggering here on Earth, are normal with to the dynamics of the universe. This should also highlight that the universe is not a static place, but it is teeming with dynamical celestial objects moving all over the place. From a far distance, however, it appears totally static – posed for any camera to take its picture.

3. The Anthropic Principle on the Existence of Life on Earth

The anthropic principle, at least one version of it, has successfully explained existence of intelligent life on Earth and the fine tuning required to make this possible.

"The argument can be used to explain why the conditions happen to be just right for the existence of (intelligent) life on the Earth at the present time. For if they were not just right, then we should not have found ourselves to be here now, but somewhere else, at some other appropriate time. This principle was used very effectively by Brandon Carter and Robert Dicke to resolve an issue that had puzzled physicists for a good many years. The issue concerned various striking numerical relations that are observed to hold between the physical constants (the gravitational constant, the mass of the proton, the age of the universe, etc.). A puzzling aspect of this was that some of the relations hold only at the present epoch in the Earth's history, so we appear, coincidentally, to be living at a very special time (give or take a few million years!). This was later explained, by Carter and Dicke, by the fact that this epoch coincided with the lifetime of what are called main-sequence stars, such as the Sun. At any other epoch, the argument ran, there would be no intelligent life around to measure the physical constants in question—so the coincidence had to hold, simply because there would be intelligent life around only at the particular time that the coincidence did hold!"

—Professor Roger Penrose
(Penrose, 2016, Chapter 10)

4. The Origins of Life on Earth: How Did Life on Earth Begin?

It has long been debated how Earth, just one of trillions of planets in our solar system, managed to cultivate life. There is evidence to suggest that our existence can be traced back to a meteorite landing on our planet billions of years ago. This theory clams that the core materials that make up DNA were transported to Earth from space via a meteorite around 3.5 billion years ago, before life on earth, as we know it, began [1]. During this period, the Earth would have been fiery and constantly bombarded by meteorites and comets from a chaotic and formative solar system. Recent analysis of three meteorites by a group in Japan has found evidence of all four of the components of DNA, namely adenine (A), thymine (T), guanine (G) and Cytosine (C) as well as Uracil, which is found in RNA [2]. But how did the presence of these chemicals give rise to living, thinking organisms?

Which came first – DNA or proteins? Life consists of the interplay between two kinds of chemicals: DNA and proteins; but which came first?

The human genome comprises approximately 6 billion pieces (or base pairs) of DNA, which come packaged in twenty-three separate chromosomes. The bases are adenine (A), thymine (T), guanine (G) and Cytosine (C). DNA has a double-helix structure and bases on opposite strands pair specifically; an A always pairs with a T, and a C always pairs with a G. There are between 30,000-80,000 genes in the human genome. Genes encodes proteins, which are the building blocks of cells and organisms. The DNA is packaged into chromosomes, which are located inside the nucleus of a cell. If a person was to walk along the spiral of DNA that they contained when they were born, they may not get to the end of it within their lifetime.

On the other hand, proteins are large biomolecules that comprise long chains of amino acids. They perform a vast array of functions within organisms, included DNA replication, metabolism and providing structure to cells and organisms.

RNA is like DNA except that it is single-stranded and contains Uracil (U) instead of thymine. Its role is to carry the message from DNA to proteins called ribosomes for translation and the formation of new proteins.

There is growing evidence that RNA came first, that is, before DNA and proteins. Many scientists believe that around four billion years ago, soon after the beginning of the earth's very existence, the presence of RNA-containing 'organisms' existed [3]. One of the problems with RNA is that it is very unstable. Therefore, it would have difficulty surviving in the high temperatures that were around at this time. To overcome these difficulties, it has been speculated that through a process of trial and error, a more stable version of RNA, DNA, emerged. In addition to this, a simple way of making RNA copies from it also arose. Hence there came into being, a primitive living organism that stored its genetic code on DNA, made its working machines of protein and used RNA to bridge the gap between them. This creature was given the name 'Last Universal Common Ancestor' or LUCA. More detailed analyses of the origins of life are given in references [1] and [4].

5. The Fermi Paradox & Drake Equation

Our galaxy, the Milky Way, is unimaginable enormous spanning a diameter of almost 2 million light-years (Croswell, 2020). It is approximately 13.5 billion years old and contains hundreds of billions of stars and planets (Milky Way, 2020). Imagine how many planets there are in the Milky Way alone that have the same conditions for life, just like here on Earth. Any reasonable person, knowing this information, might conclude that if life could start on our planet, it could start on any other similar planet. If this is the case then why haven't we detected or found signs of life (intelligent, or otherwise) in the Milky Way? Why haven't we been visited by highly advanced alien life forms?

"*Where is everybody?*", Enrico Fermi (Overbye, 2015)

The brilliant physicist, Enrico Fermi, posed this question more than seven decades ago. In that time, scientists have probed our solar system, and monitored many thousands of stars and planets for intelligent signals, see (Search for extraterrestrial intelligence, 2022). However, all these endeavors have led to nothing. Why is this a problem, you may ask. Well, Dennis Overbye explains Fermi's paradox as follows.

> *"Aliens could spread throughout the Milky Way in far less than the 10 billion years that the galaxy has been around, even if they travel at far less than the speed of light, Einstein's cosmic speed limit. Suppose, for example, that the first alien starship that sets out for another star takes a million years to arrive there and become established. Then the new colony and the original civilization each send a spaceship on a similar voyage, and so forth, so that a wave of new colonies is formed, doubling the total number every million years.*
>
> *After 10 million years, there will be 1,023 alien settlements, plus the original. After 20 million years, there will be a million. After 40 million years, if they keep it up, there would be a trillion — more than there are stars in the galaxy. By now, after 10 billion years, if there were more than one space faring civilization in the galaxy, they would be tripping over one another or one another's artifacts. But in fact there seems to be nothing."*
>
> – Dennis Overbye (Overbye , 2003)

So, the big question is, indeed, where is everyone! It would seem like, if there is intelligent life out there, something is inhibiting their ability to explore other planets in the galaxy, (Webb, 2015). The general consensus is that there is some kind of a filter that is preventing advanced civilizations exploring nearby inhabitable planets to initiate the spread of life throughout the Milky Way, as described (Overbye , 2003).

Surely, you may ask, there must be some kind of intelligent life in the Milky Way. We cannot be alone, in total isolation, marooned within an enormous galaxy island. What is the probability of communicating with an advanced civilization, you may well demand! Well, it turns out that scientists have developed a crude way of calculating the probability of the number of civilizations with which communication might be possible within the Milky Way. It is named after the scientist who first posed it, Frank Drake, the so-called Drake equation. The details are not important here, but early estimates found there to be 100 million worlds where life has been forged by evolution (Drake Equation, 2022). However, more recently the estimate has been shown to vary wildly from something near zero to just over 15 million depending on what data is used.

Until there is evidence to the contrary, from a scientific perspective at least, it looks like we are alone in our universe. Only you can decide which conclusion is more daunting; that we are alone or there are other intelligent life forms in our universe.

Finally, to close this chapter, there remain two other significant issues worth mentioning. A discussion about consciousness (as no one knows how this works, or if it even exists) and the so-called 'fifth dimension', which may be needed to explain the concept of "spooky action at a distance" as described by Albert Einstein.

BASIC PHYSICS

CHAPTER 2
OUR UNIVERSE

Our universe is the oldest natural physical entity that we, as humans, can comprehend. According to the latest scientific estimates, it is over 13.7 billion years old, (Valcin et al., 2020). Nothing in our lives can be older than the age of the universe. One may ask, what is 'age' really. Intuitively, and quite naturally, we think of this concept as a period of time. This leads to the natural question; what is time. There is no definitive answer to this question and there are as many theories on this topic as there are opinions. The fact that we cannot even agree on the concept of time is remarkable within itself. Think about what time really means. How would you come to terms with it? Well, it must start at some point in 'time' in the past. When was this time of origin? Scientists strongly believe that our universe came into existence at an event called the Big Bang, see (Big Bang, 2022) and (James et al., 2009). We will consider this remarkable event in more detail later but think of it as a 'point' in space and time. So, imagining that time started from the Big Bang doesn't tell us what it is, per se, it just tells us when it started. We have just discovered two amazing facts about our incredibly special universe: it is the oldest natural entity, and it began with the Big Bang.

Notice that we used the word 'space', together with 'time', when referring to the Big Bang event. The relationship between these two fundamental concepts turns out to be of great importance. So much so, that we will explore them in great detail in subsequent chapters. Thinking about position of the creation of our universe, or the location of the Big Bang event, is a mind twisting conundrum. What we know from current theory is that it occurred at a 'point' in space. This point, by its very definition, was an infinitesimally tiny area. It was not the size of a football, nor pinhead, not even an atom! It was literally a point in space, formally called a 'singular point' or singularity, see (Geroch, 1968). Let us think about what has been said, the entire observable universe sprang out into existence from a 'singular point' in space. Take a moment to ponder this extraordinary statement.

When measuring quantities like distance we rely on time. This is simply because the information we need to make the measurement in the first-place travels at a finite speed, called the speed of light, see (Ellis, 2005). This speed is approximately 300,000 kilometers per second, a staggering speed in terms of everyday life here on Earth. However, not that impressive when considering the scale of the universe. In fact, traveling at the speed of light for one year will only cover a distance of approximately 9.5 billion kilometers. Again, this may seem an enormous distance for us here on Earth, but it is a short distance on the scale of the universe. For example, the Andromeda galaxy, (Andromeda Galaxy, 2022) is the closest galaxy to our Milky Way, (Ferris, 1988), but it is approximately 2.5 million lightyears away. Notice we used the term 'lightyears', this is defined precisely as above, the distance covered by traveling at the speed of light for one year.

The simple act of measuring distances is complicated by the nature of the universe itself. This is because our universe is evolving. In total contrast, here on Earth, things appear static and unchanging. For instance, when you are deciding on a new table for your dining room, you take measurements of the room and then the table before purchase. What you are really doing is assuming that the dimensions of both the dining room and table do not change in time. Indeed, this is a good assumption here on Earth. This would be a bad assumption for the universe because the measurements of space itself are evolving. That is to say, the distance from the Earth to the Andromeda galaxy will change in the future because the space in between is expanding. This would also be true when measuring the distance from the Andromeda galaxy to the spiral arms of the galaxy NGC 3318, see (Figure 1).

Figure 1: The spiral arms of the galaxy NGC 3318. Image credit: ESA/Hubble & NASA, European Southern Observatory (ESO), R. J. Foley; Acknowledgment: R. Colombari. https://www.nasa.gov/image-feature/goddard/2022/hubble-sights-a-sail-of-stars/

It would seem that our universe created the notation of space and time at the Big Bang. What existed before this event we simply cannot say, with any degree of certainty, it seems to have been 'nothing', see (Bojowald, 2007). When you think of nothing, ironically, you are thinking of something. Most people, when asked to think of nothing, imagine empty outer space or a vacuum, (Stoner et al., 2012), of some kind. But even the empty space, say between the Earth and the Moon, has a zoo of sub-atomic particles. These particles are the building blocks of matter and energy of our universe. There are even so-called virtual particles that come in and out of existence that fill the vacuum of empty space (we will dive into more detail about these types of particles later). If you think you can achieve nothingness by sucking all the air out of a strong container, then you would be very disappointed. Any vacuum chamber on Earth would be plagued with sub-atomic particles and virtual particles just like the vacuum of outer space, (Boyer, 1985). So, what is 'nothing'? Whatever it is, our universe is expanding in it at a faster and faster rate, see (Carroll, 2004).

There are other mysteries to be solved like where the matter came from to form stars, planet, galaxies and so on, that we observe today, (Krauss, 2012).

Why weren't there just photons, which are particles of light, emitted at the Big Bang. After all, if you think of an explosion, like a firecracker exploding, you generally see flashes of light and a loud crack. What you do not see, nor expect, are large clumps of matter to be produced. So, why do we live in a matter-dominated universe? A related mystery involves so-called anti-matter, see (Close, 2009). It is well-known among scientists that every particle has an anti-particle. For example, an electron has an anti-electron called a positron, they are total opposites in terms of charge. An electron has a negative charge, -1, while a positron is positively charged, +1. They are identical in every other aspect, such as their mass. It is important to understand that if an electron and positron collide, then they will annihilate each other and produce photons (particles of light). Atoms are widely known to be composed of protons, neutrons and electrons. Scientists now understand that protons and neutrons too are composed of 'fundamental particles. These particles, and their anti-matter counterparts, belong to a family of particles called Leptons and are therefore considered to be fundamental particles (building blocks of matter and radiation). So, shouldn't there be as much, so-called, anti-matter as there is matter. Why do we only see matter composed of protons, neutrons and electrons? These are important questions that must be answered by any theory of our universe.

You may well ask, if matter and anti-matter always annihilate each other then how can this lead to the creation of matter? At the ultra-high temperatures of the Big Bang, and according to theoretical calculations, it is possible that for every billion positrons, one billion and one electrons were created. A similar scenario would exist for protons and neutrons, and their anti-matter counterparts. Such an excess, although minute, would be crucially significant. The single unpaired proton, neutron and electron could combine to become the material that formed matter in the universe. According to this theory, the universe is built out of a tiny residue of unbalanced matter that survives as a relic of the first brief moment of existence. Thus, the appearance of matter without antimatter can be explained on ordinary physical grounds, (Canetti et al., 2012).

In some ways, we are fortunate that there is an absence of anti-matter in our universe. If there was just as much anti-matter as there is matter then, eventually, they would totally annihilate each other to produce radiation, including visible light. Our universe would be filled with uniformly distributed radiation that would continue to expand after the Big Bang. This radiation would cool off as this expansion process continues without producing any matter. If there is no matter in the universe then gravitational processes that help form galaxies, planet and stars would not occur. This would be a universe unable to bear life, as we know it at least, and would be cold, dark and lifeless.

Yet another mystery involves how smooth and flat our universe is today. This tells us that it would have to be much smoother and flatter just after the Big Bang. The problem with this idea is that vastly different regions of the expanding early universe would have to be in contact with each other. Put more scientifically, these regions would have to be in causal contact. This would require speeds faster than light, which have never been measured experimentally. To better understand this issue, think of a closed bottle containing a gas of some kind. If the bottle were to be opened in a sealed room, then it would occupy the volume of the room. This is exactly what you would expect, as all the gas particles are essentially in causal contact. The walls of the room define the boundaries and the force of the particles colliding with the walls (including floor and ceiling) can be thought of as the pressure of the gas. A particle of the gas hitting one wall can be in causal contact with another gas particle on the other side of the room through gas-particle to gas-particle collisions. There may be many, many, such collisions in this 'communication' process. The result will be a smooth distribution of gas throughout the room.

Now think of the walls of the room expanding out so that the volume is ever-increasing. If you release two identical, but separated, bottles of gas that are also rapidly moving away from each other then how can they 'communicate'? There can be no collisions between particles from each bottle because they are so far away. Suppose the speed of separation of the bottles is far greater than the average speed of the gas particles. The two clumps of expanding gas will not be in causal contact with each other. This will result in a clumpy distribution of gas in the expanding room and not a smooth distribution. This is a good analogy for the processes that scientists use to envisage the Big Bang.

This analogy leads to a remarkably profound implication for our universe. There must have been some extraordinary process that helped smooth out the distribution of matter and, at the same time, kept the space exceedingly flat.

This process is the much-celebrated cosmological inflation, see (Inflation (cosmology), 2022). It was introduced in the 1980s by a group of cosmologists and physicists. Most notably by Alan Guth and Andrei Linde, but also a host of others. The general idea of inflation is that the early universe experienced an exponential expansion, between 10^{-36} seconds after the Big Bang to around 10^{-32} seconds. Extraordinarily, this expansion precisely explains the large-scale smoothness of our universe and a host of other unanswered questions. This is a simple idea that results in the irregularities of the early universe being expanded so rapidly, over great distances, that they are 'stretched' out or smoothed over.

There are five stages of the evolution of our universe to date, see (Chronology of the universe, 2022). They are: the very early universe, the early universe, the Dark Ages and large-scale structure emergence, the current state of the universe and the future of the universe. The very early universe takes place over the first pico-second of the big bang, which is a staggeringly short 10^{-12} seconds. You may not think a lot can happen within this absurdly brief period of time, but it can. This is called the Electroweak epoch, and it is when the laws of physics are established. This epoch is preceded by three others, namely the Planck epoch, the grand unification epoch, and the inflationary epoch, in that order. The radiation temperature, by the end of the Electroweak epoch, was a colossal 1 quadrillion Kelvin (this is 1 followed by 15 zeros!).

The Early Universe.

The mass of the early universe, as we understand it, had not been formed. If there was no mass around, then there could not have been a timescale. This can be demonstrated by considering two of the most famous formulas in physics. One is Einstein's equation relating energy, denoted by E, to mass, denoted by m,

$$E = mc^2$$

where the quantity c represents the speed of light. The other is Planck's equation relating energy, also denoted by E, to frequency, denoted by v,

$$E = hv$$

where h is the Planck constant. One can now combine these equations by equating the energy terms on the left-hand sides. This produces the relation,

$$m = \frac{h}{c^2} \nu$$

where h/c^2 is just a constant factor, and we have divided both sides of the equation by the speed of light squared. This relationship implies that mass has an associate timescale because mass is proportional to frequency. The frequency defines the timescale. Interpreted another way, this relationship also implies that massless particles, like photons which are particles of light, have no associated timescale. In this sense, massless particles like photons are eternal. You should understand that massless particles like photons will always be massless particles.

Given the relationship between mass and time, one may conclude that because there was no mass in the early universe, there was no relatable timescale. The process that endows particles with mass, that are meant to have mass, is the so-called Higgs field. This field permeates all space, and the particles of the field are so-called Higgs bosons, see (Higgs boson, 2022). As particles interact with the Higgs field they get encoded with mass. This is a profoundly important piece of theory that was developed by teams of scientists dating back to the mid 1960s. It was verified by the Large Hadron Collider experiments of 2012 and 2013, see (O'Luanaigh, 2013). The success of the theory surfaced to mainstream media, where the Higgs boson has been popularized as the 'God particle' and even found its way into Hollywood.

The early universe lasted around 370,000 years, see (Kolb et al., 2018). Within the first second of this stage, various kinds of subatomic particles were formed, and almost equal parts of matter and anti-matter were produced. As expected, they annihilated each other, and as stated above, a resultant amount of the matter we see today persisted. Most importantly, after the one-second mark, a host of sub-atomic processes occur. This gives rise to the possibility of forming so-called Primordial Black Holes, see (Primordial black hole, 2022). A further twenty minutes and the universe has cooled so that it is no longer hot enough for nuclear fusion, see (McCracken et al., 2005). It is still, however, far too hot for neutral atoms to exist or even photons to venture far. If somehow you were alive during this period, then it would be an opaque world indeed. The recombination epoch begins at around 18,000 years when sub-atomic particles start to combine in various forms. Up until this time, it is a universe dominated by the radiation era. Then, around 47,000 years, the universe starts to cool, and its behavior starts to change radically to one dominated by matter.

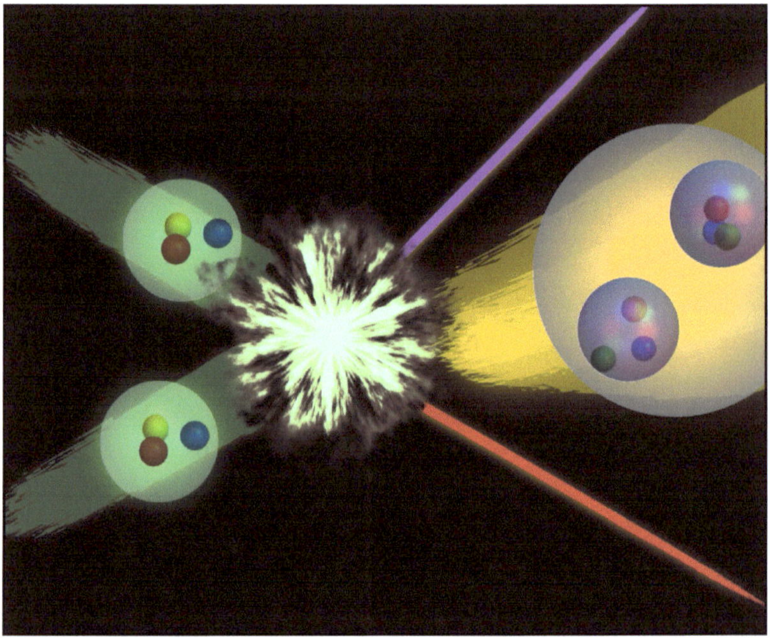

Figure 2: Nuclear physicists are using the supercomputer, Titan, at the Oak Ridge Leadership Computing Facility (OLCF) to study particle interactions important to energy production in the Sun and stars . Image Credit: https://commons.wikimedia.org/wiki/File:Particle_ Interactions_ Calculated_on_Titan_Support_the_Search_for_New_Physics_Discoveries_25373768837.png

Around 370,000 years neutral hydrogen atoms, common to us today, finally formed (see, Figure 2.). This means that light can reach much farther and so the universe became transparent for the first time, see (Weinberg, 1994). Remarkably, this light (or the particles of this light called photons) can be detected today from a knowledge of the cosmic microwave background. This is what is formally called the observable universe. In other words, the oldest possible observation we can currently make of the universe (see, Illustration 1.).

Illustration 1: Distribution of dark matter, galaxies, and hot gas in the core of the merging galaxy cluster Abell 520. Image Credit: NASA, ESA, CFHT, CXO, M.J. Jee (University of California, Davis), and A. Mahdavi (San Francisco State University). https://www.nasa.gov/multimedia/imagegallery/image_feature_2189.html

The Dark Ages and Large Scale-structure

The so-called dark ages lasted over a period from 370,000 years until about 1 billion years, see (Loeb, 2006). As the name suggests, this period saw the universe becoming darker and darker due to the ever-decreasing density of photons in the visible spectrum (simply put, there was little light!). At a later time, the earliest stars and galaxies started to form due to gravitational attraction. Scientists have not been able to determine exactly when the earliest stars were formed but they are confident it occurred within this time period. By the end of the dark ages, around 1 billion years after the Big Bang, the universe started to take on the large-scale structure that we see today.

The Current Universe

At the end of the dark ages, 1 billion years ago, to the current universe not a lot has changed on a large scale, see (Carroll, 2006). Our Milky Way formed around 9 billion years ago, and life emerged on Earth after around a billion years. On a large scale, scientist expect the universe to look pretty much the same for billions of years to come.

There are two competing forces at play in the evolution of the universe. On the one hand, there is the expansion of the universe itself, which is repelling galaxies away from each other. On the other hand, this is rivaled by gravitational attraction, that is bringing massive celestial objects closer to each other. Scientists believe that around 5 billion years ago the expansion of the universe started to win over gravitational attraction, see (Overbye, 2017). This victory seems to be getting stronger as time goes on because it is not only expanding but also accelerating. This is a profound discovery, and we will revisit this result in later chapters.

The Future Universe

Of course, the last stage, the future of the universe is unknown, and we can only be guided by theoretical predictions and, perhaps, speculation. One of the most profound mysteries of our understanding of the universe is the missing matter and energy problem, commonly known as dark matter and dark energy, see (Dark matter, 2022). The 'dark' in their name refers to the fact that they only interact with the universe through gravitational means alone. They cannot be observed, measured directly or indirectly, and their composition is unknown, see (Dorminey, 2021). Most frighteningly, together, they make up approximately 85% of the matter and energy in the universe. We will dive into more detail about this fascinating puzzle shortly.

CHAPTER 3
BASIC PARTICLE PHYSICS

It would be a good idea, at this stage, to briefly mention some useful facts about fundamental particle physics, see (Particle physics, 2022). We mentioned some fundamental particles previously in our discussion above. The main thrust of this book, of course, is focused on the large scale. However, a discussion of fundamental particle physics, that explains the building blocks of matter, is paramount for a clearer understanding, see (Larkoski, 2019).

There are three general types of elementary particles. Recall, by elementary, we mean they are building blocks for more complex structures like atoms and molecules. Think about a house: its building blocks are bricks, planks of wood, nails, etc. The three building blocks of matter and energy are leptons (green), quarks (purple) and bosons (orange and yellow), see Table 1. Quarks are not generally found on their own but are the building blocks for hadrons (either baryons or mesons), (see, Figure 3.). Moreover, leptons and quarks are fermions, see (Fermion, 2022), while mesons and force-carrying particles are bosons, see (Boson, 2022). Fermions obey Fermi–Dirac statistics and bosons obey Bose–Einstein statistics, see (Lighthouse, 2014). This can all sound overwhelming at first, but we are just categorizing them like a shopping list, veggies, poultry, dairy, etc.

There is a so-called standard model of particle physics that categorizes all the various types of particles in our universe. The elementary particles that make up the standard model are displayed in Table 1. By elementary particles, we mean they are the things we put together to form an atom of stuff, e.g., water, dust, rock, and so on, in the universe. For example, consider a water molecule, it is made up of two atoms of hydrogen and one atom of oxygen (see, Figure 4.). But the hydrogen and oxygen atoms are not themselves 'elementary' because they are made up of other particles.

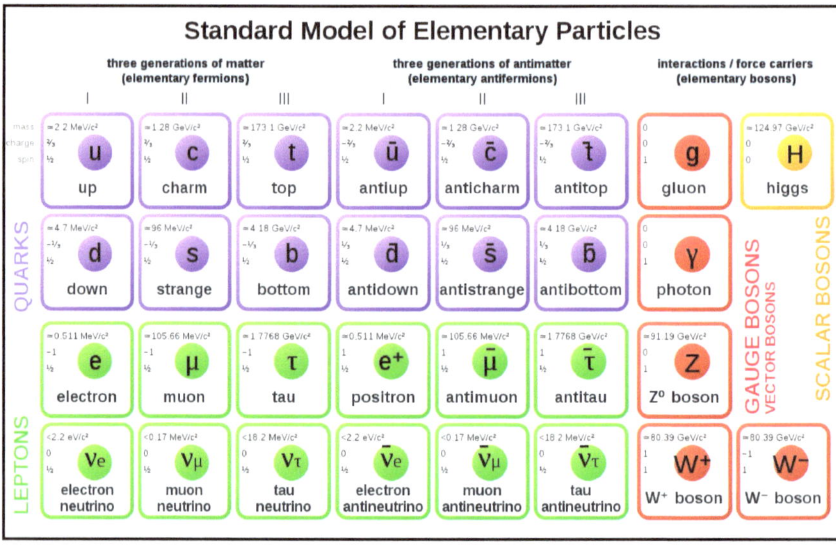

Table 1: Standard Model of Elementary Particles consisting of six quarks (purple) and six leptons (green), and their anti-particles. In addition to gauge and vector bosons (red) with the Higgs particle as the only scalar boson (yellow). https://en.wikipedia.org/wiki/File:Standard_Model_of_Elementary_Particles_Anti.svg

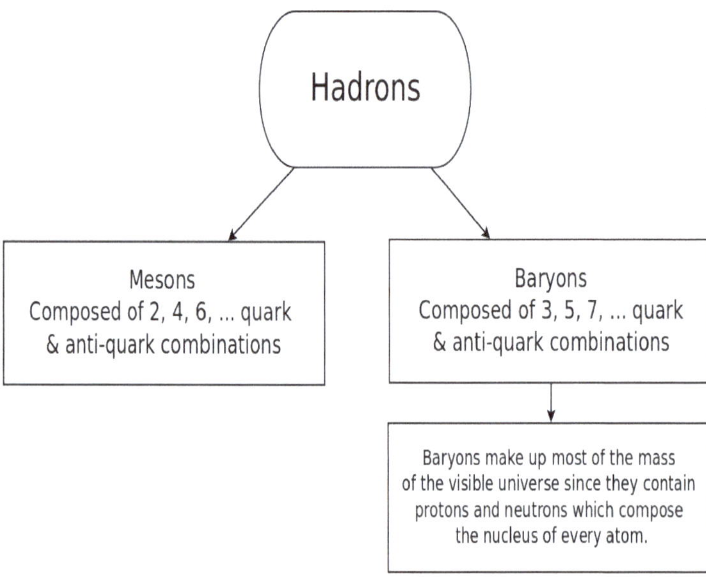

Figure 3: The hadron family of elementary particles. Mesons and baryons belong to the hadron family. Mesons are composed of an even number of quarks in quark & anti-quark combinations. Baryons are composed of an odd number of quark & anti-quark combinations. Baryons make up most viable matter in the universe.

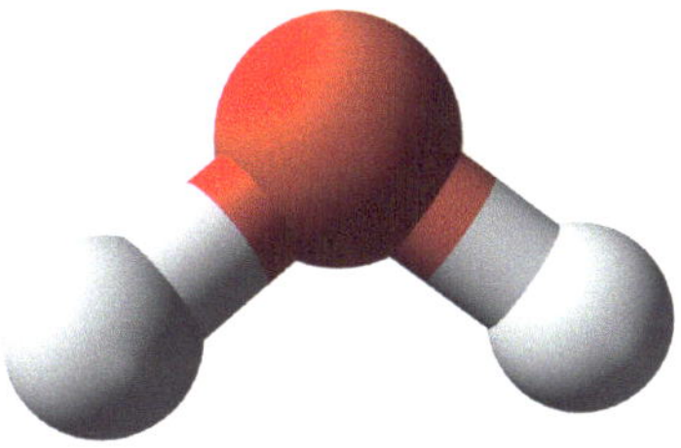

Figure 4: Model of a water molecule. The two hydrogen atoms are represented in white while the oxygen atom in red. Image credit: https://en.wikipedia.org/wiki/File:Water-3D-balls-A.png

Let us think about the hydrogen atom, (Gregory, 1972), it is composed of a neutron and proton at its nucleus with a single electron orbiting around. At least, this is the classic picture most people have been taught in school. The electron belongs to a family of elementary particles called leptons (see Table 1., green tiles). Therefore, it is an elementary particle since it is not composed of any other known particles. On the other hand, a proton is not an elementary particle because it is composed of so-called quarks. Quarks are fascinating elementary particles because they cannot be found in nature in isolation. They must be associated with a particle like a proton to maintain stability. Going back to our example, the three quarks that make up a proton are two 'up' and one 'down', (see, Figure 5.). A Proton is a composite particle made up of three quarks and therefore belongs to the so-called hadron family, (see, Figure 3.). All members of the hadron family are composite particles made up of at least two quarks. Finally, the neutron is also made up of three quarks, just like the proton but with two 'down' and only one 'up', (see, Figure 6.).

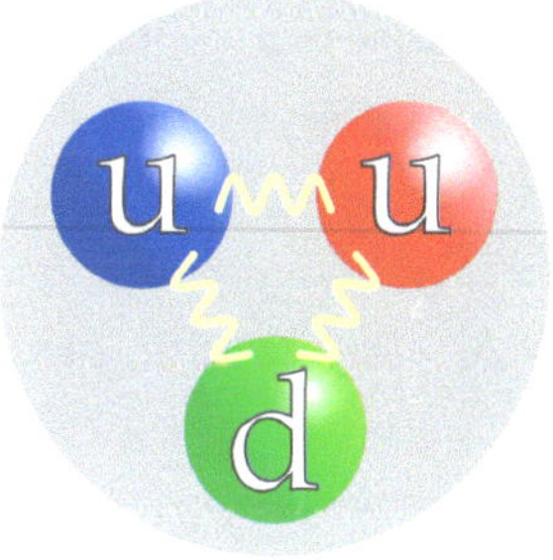

Figure 5: Quark structure of a proton. There are two 'up' quarks and one 'down' quark. Image credit: https://en.wikipedia.org/wiki/File:Quark_structure_proton.svg

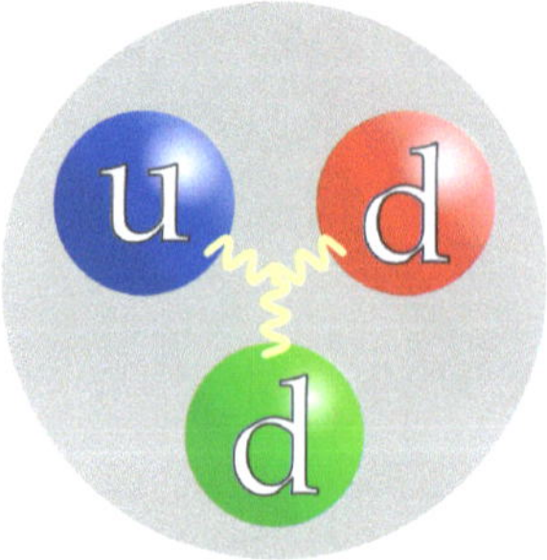

Figure 6: Quark structure of a neutron. There are two 'down' quarks and one 'up' quark, in contrast to the case for the proton. Image credit: https://upload.wikimedia.org/wikipedia/commons/thumb/b/bf/Neutron_quark_structure.svg/1024px-Neutron_quark_structure.svg.png

It should be clear that baryons are composite particles, composed of at least three quarks. They are still, however, described as 'sub-atomic' despite being a composite particle. The number of quarks involved in the composition of the baryon must be odd. This odd number must start from three quarks, or can include five, seven, nine and so on. As such, baryons belong to the hadron family of particles, see (Hadron, 2022). On the other hand, mesons also belong to the hadron family, however they must be composed of an even number of quarks. The minimum number of quarks that make up a meson is two. They can also be composed of four, six, eight, and so on, quarks.

We should mention that every particle has an anti-particle. This is true for quarks and leptons. A close inspection of Table 1 clearly shows that all six quarks are matched by their corresponding anti-quarks. For example, an 'up' quark has a corresponding anti-particle called the 'antiup' quark. This is also true for leptons, for example, an electron has a corresponding anti-particle called the positron. If the two ever come in contact with each other, they will be totally annihilated. This is a very important concept in understanding the workings of the universe. Also, particles in general have three properties: charge, spin and mass. You can see these properties displayed in small font in the top left of every square representing a particle in Table 1. These properties are used in their classification, as will be discussed below, see (Tully, 2011).

The first of these properties is the mass of a particle. You may think that this is self-evident, and it is for the most part, but how can we think of a particle with no mass at all, i.e., zero mass. For example, a photon has zero mass, see

Table 1., under bosons (orange). How can something with zero mass even exist? If your car had zero mass then there wouldn't be anything in front of you, as you could imagine. However, extraordinary things occur on the extreme scale in the domain of elementary particles. The photon simply exists as energy and not mass. We know from Einstein's famous formula, as discussed previously that relates energy with mass, that the two concepts are interchangeable. From this perspective, having a elementary particle with zero mass is not surprising at all. In fact, it is a fundamentally important property to have, as we will see later.

So, we can understand the concept of mass intuitively, but what does it mean for an elementary particle to have charge? This can be a strange concept because when you think of something being 'charged' you most commonly think of, say, a battery. A battery is composed of many specialized types of materials and, in some cases, liquids. On the other hand, an elementary particle is quite literally the smallest building block of its type. It is only in rare cases that an object can have an overall charge. In the case of elementary particles, this is one of the fundamental ways atoms are formed.

To better understand these phenomena, let us revisit the hydrogen atom. Recall, it has one proton and neutron making up its nucleus, and an electron orbiting around this nucleus. The overall charge of the nucleus is positive, since the proton is positively charged. The neutron, as you may guess from its name, has no charge or it's neutral. The electron orbiting around this nucleus is negatively charged. This means it is attracted to the positive charge of the nucleus since positive and negative charges attract, whilst positive and positive, or negative and negative charges repel. The reason the electron does not crash into the nucleus is because it is in a stable orbit, or in one of the so-called stationary states, see (Hydrogen atom, 2022).

So, there are three possibilities when it comes to charge, an elementary particle can have one unit of positive charge, +1, be neutral, 0, or have one unit of negative charge, -1. This is clearly an integral part of Table 1., for leptons. But, what about quarks, they appear to have fractional charge. Remember, we said that quarks must always form a particle, like a proton, and are not found in isolation. Therefore, when they combine the overall charge is either -1, 0 or +1. Consider the proton as an example, we learned that a proton is made up of two 'up' quarks and one 'down' quark, (see, Figure 5.). The charge of an 'up' quarks is 2/3, and the charge of a 'down' quark is 1/3.

This means that the charge of a proton must be (+2/3) + (+2/3) + (-1/3) = +1, as expected. What about a neutron atom, it is composed of one 'up' quark and two 'down' quarks. So, its charge must be (+2/3) + (-1/3) + (-1/3) = 0, again, as expected, see (O'Luanaigh, 2014) and (Gell-Mann, 2010).

The spin of an elementary particle makes up the three properties used to classify them into distinct families. A way to image the concept of spin is to think of a figure skater spinning about an axis that goes from her head to the tip of the skate touching the ground. She makes a revolution about this axis but is not moving laterally. An elementary particle does this without the need for skates! More formally, one speaks of the intrinsic form of angular momentum (spinning) as a fundamental property of elementary particles.

An inspection of Table 1., shows that all fermions (leptons and quarks) have spin ½. On the other hand, bosons have either spin 1 or zero. Therefore elementary particles are said to have either half-integer spin, which is called fermion statistics, or integral spin, boson statistics. So, for example, an electron has spin ½ and therefore it must follow fermion statistics. In contrast, a photon has spin 1 and must follow boson statistics, see (Cottingham et al., 2001).

An extremely important and relevant particle, regarding the pursuit of understanding the universe, is the so-called force-carrying photon, see Table 1 (orange). It turns out that for electromagnetic fields, the force-carrying particle is the photon. You should know that each and every field in physics has an associated force-carrying particle, see (Pillitteri, 2017). In the case of an electromagnetic field, it is a photon. This makes sense because a photon is associated with light, for example, the light from our Sun. It turns out that this association plays a fundamental role in cosmology, as we will see later. Another important force-carrying particle is the Higgs boson, see (Higgs boson, 2022), which also plays a fundamental role in the creation of the universe by endowing elementary particles with mass. The Higgs field is a rather strange thing in physics and not fully understood, at the time of writing this book. A good analogy to understand the general idea, is to think about dropping peanuts into a bowl full of (Bee's) honey - the peanuts represent the particles and the honey the Higgs field. The peanuts would emerge from the honey heavier because they have been coated by the experience of moving through the viscous honey. That is, the honey would stick to the peanuts making them heavier.

The most intriguing mystery in modern cosmology is determining the particle composition of dark matter, see (Bertone, 2018). It cannot be composed totally of baryonic matter because baryons contribute to most of the ordinary matter (or atoms) in the universe. We should point out that protons and neutrons belong to the baryonic family of particles. Since most common matter is composed of these particles, we are justified in ruling out ordinary matter as being the main constituents of dark matter. The main reason being that baryonic matter is easily detectable and dark matter has so far evaded any type of detection.

This means that dark matter must be composed of non-baryonic matter that cannot be detected through any known scientific means, see (Del Popolo, 2013). That is, sub-atomic particles that are neither protons nor neutrons. So, for example, not like matter we encounter here on Earth or in our Milky Way, (see, Figure 7.). This makes the challenge of detecting dark matter ever so hard or near impossible, which is why it has puzzled scientists for many years.

The Wilkinson Microwave Anisotropy Probe (WMAP), see (Wilkinson Microwave Anisotropy Probe, 2022), has played a major role in the pursuit to understand dark matter's composition, (see, Illustration 2). One of its many successes was to determine that the total baryonic matter (ordinary atoms) of the universe actually only makes up around 5% of the universe. WMAP also established that dark matter accounts for 23% of the total matter and energy content of the universe. It is hoped that WMAP will go one step further to uncover solve the puzzle of dark matter. We will consider this mystery in more detail later.

Figure 7: Our Milky Way galaxy shows a mishmash of gas, charged particles and several types of dust. Image credit: ESA/NASA/JPL-Caltech. https://www.nasa.gov/jpl/planck/pia18912

Illustration 2: The Wilkinson Microwave Anisotropy Probe (WMAP) used the moons gravitational field to slingshot to a location in space called the L2 Lagrange point. It covered a 1.5 million kilometer journey from the Earth. Image Credit: NASA / WMAP Science Team. https://wmap.gsfc.nasa.gov/media/990387/990387_640.jpg

CHAPTER 4

BASICS OF THERMODYNAMICS: ENTROPY

An area of physics called thermodynamics is one of the most important theories in science, see (Thermodynamics, 2022). Its initial development dates to the mid-15th century. It is fundamental in explaining, for example, all phenomena that involve heat exchange. One of the most intriguing concepts in thermodynamics is called entropy, see (Entropy (classical thermodynamics), 2022). Basically, this means that everything in our universe moves from order to disorder. One way we can think about this is to consider the analogy of a gas being released within an empty room. The idea is that this gas is an ordered system because it is contained in a sealed jar. Now, this jar is taken to a corner of the room and released. Over time, it would reach every part of the room's empty space. Put another way – the gas would occupy the volume it is given. So, you can think about the gas in the jar as an order system, when released into the empty room it becomes disordered, see (Second law of thermodynamics, 2022).

The universe, including everything that happens in it, is considered an isolated system. What this implies is that entropy always increases and never decreases, see (Ribeiro et al., 2021). Our universe is moving from an ordered state to a disordered one. To better understand this, consider your favorite beverage in a tall glass, with some added cubes of ice. You leave your drink on a table, in a room, of your house. What will happen over time is that the ambient temperature of the room, in your house, will equalize with the temperature of your drink. Heat, or thermal energy, from the warm air in your house and room will move to the cooler drink containing ice cubes. Over time, the temperature of the drink will increase while the surrounding temperature of the room will decrease slightly. Of course, we are assuming the house temperature is isolated from the outside environment. It is only a matter of time before the temperature of your drink and the temperature of the room become equal. When this occurs, we say that the isolated system (the house and your drink, see Figure 8.) is in thermal equilibrium.

Figure 8: Glass of water with ice cubes and lemon. The temperature of the drink increases to match its surroundings. Image credit: https://commons.wikimedia.org/wiki/File:Ice_water_with_lemon.jpg

Let us now inspect what has happened to the entropy of our thought experiment. The ambient temperature of the house, and room where the drink was placed, decreased over this duration to thermal equilibrium. This means that the thermal energy decreased because some of it was transferred to the drink, so the entropy of the house also decreased. On the other hand, the drink itself increased in thermal energy because it was initially colder than the ambient temperature of the house. So, the entropy of the drink actually increased. However, the entropy of the drink has increased more than the entropy of the house has decreased. In other words, the net change in entropy has increased. Intuitively, you can convince yourself that this will always be the case because you generally can't perceive the temperature change of a room or house when you have a cold drink. But you certainly know when your favorite cold beverage turns warm to the touch! Of course, this is not scientific because you could have started with a hot cup of coffee instead. But the result of increased entropy would also hold.

The results of the above are a good analogy for our universe. When processes in the universe consist of systems that reach thermal equilibrium, the entropy will undergo a change, from its value when it started, to its maximum value. One could then view entropy, of the thermodynamic system, as a measure of how far the equalization process towards thermal equilibrium has progressed. With respect to our experiment, when the drink with ice cubes melted, is at the same temperature as the house we know the entropy is at its highest. So, the entropy of the universe will always increase as it continues to expand and endeavor to reach thermal equilibrium, see (Gibbs, 2012).

The ideas behind entropy apply to our universe because it is considered to be an isolated system. According to the so-called 'second law of thermodynamics', see (Second law of thermodynamics, 2022), the entropy of the universe, or any other isolated system, will always achieve a state of thermodynamic equilibrium. The entropy at this equilibrium stage will be at its highest and can never decrease with time. A remarkable thought now comes to fruition: if everything behaves in the universe in this way, why couldn't we think of entropy as a concept of time? After all, it seems like they share a key property of both moving strictly in one direction. Scientists describe this concept as an arrow of time, see (Entropy as an arrow of time, 2022). Generally, we don't expect to see time going backwards and neither does entropy decrease for the evolution of the universe.

As we will discover later, there is a profound and deep relationship between the concept of entropy and the physical properties of so-called Black Holes. We will dive into this interesting topic in more detail at a later stage but let us consider some broad ramifications. Basically, Black Holes are super dense objects formed by condensing matter and energy. Usually, this happens when very large stars collapse in on themselves. The gravitational attraction associated with them is so strong that not even light can escape. Once you get too close to one, your fate is sealed and there is no coming back to our universe. This point of no return is called the 'event horizon' and it is shaped like a ball, see (Event horizon, 2022). So, imagine a super compact ball-shaped object that is swallowing everything that comes near it like a cosmic vacuum cleaner that is totally black, (see, Figure 9.).

Now, imagine you are on a spacecraft enjoying a fresh cup of hot coffee. Your cup of coffee, as we now know from the discussion above, has high entropy before it cools off and conforms to thermal equilibrium in your spacecraft, an isolated system. At that point in time, your spacecraft ventures near a Black Hole. You then decide to throw your cup of coffee into the Black Hole! Thermal equilibrium was not established between the interior temperature of your spacecraft and the hot cup of coffee. Once it enters the Black Hole - that is the event horizon-entropy is lost from our universe. This means that entropy has decreased, in direct violation of the second law of thermodynamics. How can this be the case? Perhaps, in fact, Black Holes swallow entropy as they do matter and energy, thereby violating long held and tested results in thermodynamics. This was indeed a profound mystery that was solved by one of the most ingenious physicists of the late 20th century. Incredibly, it turns out that no violation of thermodynamics takes place, as we will discover in more detail later, see (Bekenstein, 1973) and (Ouellette, 2015).

Figure 9: Supermassive black holes are believed to reside in the hearts of all large galaxies. This is an artistic illustration of the local galaxy Arp 220, captured by the Hubble Space Telescope. The bright core of the galaxy is enhanced with an artistic impression of a pair of jets emanating in opposite directions. This illustration indicates that the central black hole's activity is intensifying. Image Credit: NASA/JPL-Caltech.

If the total entropy must always increase in the natural evolution of our universe, then it is fair to say it started with lower entropy. The entropy of the universe is greatly influenced by the gravitational attraction of cosmological objects. Matter tends to distribute itself in an expanding universe, much like a gas does when released from a sealed jar into a room. Initially, the gas is packed densely (under pressure) into the jar, but this changes when it is released into the room. The particles of gas explore the boundaries of the room over time. If you apply this analogy to our universe, where the particles of gas represent celestial objects that evolve to form stars and so on, the extra element that would need consideration is the role of gravity. Some of these stars may eventually evolve into Black Holes. It turns out that the entropy of a Black Hole is related to the surface area of its event horizon. Remember, the event horizon is shaped like a round ball. That is, the surface area of a Black Hole is proportional to its entropy. The bigger the Black Hole, the bigger the surface area of the event horizon, the bigger the entropy. This is important because it helps the universe increase its total entropy as it evolves and expands. Compared with any celestial object of similar size, Black Holes have the maximum possible entropy, see (Bekenstein, 1973).

In everyday life, you never see processes in reverse, such as a broken egg reassembling itself and jumping back onto a tabletop. You never see a demolished building reconstruct itself from rubble, or a person 'shrinking young' when they are old. Warm drinks do not develop ice cubes and become cold. It is not correct to say that these processes are 'impossible', but you can think of them as highly improbable. This is because systems are more inclined to evolve in the direction of increasing entropy. We can see the concept of an arrow of entropy at work here and the fact that it is naturally aligned with the arrow of time. From this perspective, we are viewing entropy from a statistical framework. Basically, the idea is that an ensemble of an event is collected, and an average is taken as the outcome. (Think of an ensemble as a collection of photos taken of each time you run an experiment or event.) On a technical level, we say that these processes are irreversible.

> *"The increase of disorder or entropy is what distinguishes the past from the future, giving a direction to time."*
>
> — *Professor Stephen Hawking,*
> A Brief History of Time

The real mystery here is why entropy actually increases for our universe in the first place. Why should the universe be so ordered at the Big Bang event? Sir Roger Penrose (Penrose, 2011) argues that the gravitational degrees of freedom (basically, the freedom that gravity must act in all directions) do not fulfill a role in the processes of the early universe. They are subsequently released as the evolution of the universe continues. When the cosmic microwave background (CMB), see (Cosmic microwave background, 2022), is analyzed it reflects high entropy which is uniform. But this contribution is only from matter, not from the curvature of spacetime. Then, as the gravitational degrees of freedom begin to unravel, the matter content starts to coalesce. It is this 'clumping' effect that leads to increased entropy.

This clumping effect is not mirrored in the cosmic microwave background (CMB), which is almost uniform, as expected. Clearly, this process leads to the formation of planets, stars and so on, and they contribute to the increase of entropy. Ultimately, the Black Holes that are produced will substantially increase the entropy of the universe. This is especially true for the largest of Black Holes, known as Supermassive Black Holes. Their mass ranges from millions to billions of solar masses. Imagine something millions to billions of times the mass of the Sun; this is simply colossal as far as mass of a celestial body is concerned.

CHAPTER 5

THEORIES OF GRAVITY AND THE UNIVERSE

Figure 10: Sir Isaac Newton. https://commons.wikimedia.org/wiki/File:Portrait_of_Sir_Isaac_Newton,_1689.jpg

There is no doubt that Sir Issac Newton, (see, Figure 10.), was a genius, whose ideas and discoveries span over many centuries. He proposed his ideas and discoveries on gravitational phenomena in the late 17th century and most of his ideas have stood the test of time. One of his theories considered the gravitational phenomena. These phenomena were not well understood at the time. He strongly believed that any laws governing the gravitational interaction between two bodies of a given mass would be expressed as a force. Each body would feel a force due to the presence of the other. The idea was that two heavenly bodies, like the Earth and the moon, attracted each other inversely with respect to the distance that separated them. In other words, if the bodies were separated a far distance, then the force would be weak, whereas it would be stronger if they were close together. This meant that the force due to gravity between two celestial bodies was expected to be weak, see (Newton's law of universal gravitation, 2022). On certain scales, Newton's law of gravitation provides accurate empirical predictions. This theory successfully predicts the motion of the moon around the Earth and the motion of the Earth around the sun. The success of the theory may be extrapolated beyond the distance between the Earth and the moon. The theory was so successful it took hundreds of years before it was eventually challenged, see (Lewis, 2016).

The Newtonian picture of the gravitational phenomena only accounts for terrestrial observations. Albeit, to a high degree of accuracy. Newton formulated his theory by taking measurements between observable, Earthly, objects. Basically, everyday objects that did not change vastly in space nor time. Objects like trains, automobiles, and falling apples, that travel at very low speed. A typical speed limit on roadways is 60 kilometers/hour, or 0.0167 kilometers/second. This is minuscule compared with the speed of light, approximately 300,000 kilometers/second. Newtonian gravity completely breaks down at, or even a fraction of, the speed of light. This is because the relationship between acceleration and gravity is not accounted for in the theory. It takes an enormous acceleration to achieve even a fraction of the speed of light. When scientists, at the turn of the 19th century realized this fact, they started searching for a new theory of gravity to account for this discrepancy, see (Deaton, 2019).

The new theory that emerged was proposed by Albert Einstein, see (Figure 11), called the General Theory of Relativity. This theory could explain the gravitational phenomena across all speeds up to and including the speed of light. The speed of light, defined by the speed of a particle of light called a photon, plays a significant role. One of the key ideas that Einstein suggested is that nothing can travel faster than the speed of light. Put another way, it was the speed limit of the cosmos. This is in total contrast to the Newtonian picture, where there is no speed limit at all. In fact, Newtonian's theory requires a rather uncomfortable concept of 'action at a distance'. Put differently, this meant that things could occur instantaneously, even if the distance separating them was vast. The success of Einstein's theory did not only mean it generalized Newton's theory but also corrected it in certain parts, see (General relativity, 2022). It has since produced profound insights and discoveries, (see, for example, Illustration 3).

Figure 11: Albert Einstein. https://commons.wikimedia.org/wiki/File:Albert_Einstein_1947.jpg

This leads to a conundrum: how can Einstein's theory of gravity be a radically different concept to Newton's theory yet lead to basically the same answer in some limiting sense? This is expected when the speeds and volumes involved are relatively small. We know that the results of Newton's calculations in these situations are highly accurate. Scientists from NASA use Newtonian laws to send people to the moon and back. Shouldn't then one expect to recover Newtonian theory from Einsteinian theory in some appropriate limit? Think of the concept of a limit as one physical or mathematical object approaching another, respectively. This is exactly what can be done by considering Einstein's theory in the limit where the speed of light approaches infinity. Remember, the maximum speed in the theory of relativity is the speed of light, whereas in Newton's theory there is no limit. Physically, assuming the actual speeds and volumes of the objects involved are terrestrial, like buses and cars, we miraculously recover Newtonian theory from Einsteinian theory.

Illustration 3: Artist's concept illustrates a supermassive black hole. Image credit: NASA/JPL-Caltech. ttps://www.nasa.gov/mission_pages/nustar/multimedia/pia16695.html

Another profound concept of Einstein's was based on earlier work of a scientist named Minkowski. He theorized that we live in a four-dimensional "spacetime". The concept of a so-called Minkowski spacetime would be the foundation of the theory of relativity (both special and general) and play a pivotal role in modeling the current state of the universe. What this means is, instead of thinking about three-dimensional space (imagine a room having length, width and height) and time separately, we join them together. This unification is called spacetime. You can think of this as literally joining the words "space" (three-dimensional space) with "time" (which is one-dimensional) to form the word "spacetime" (which is four-dimensional), see (Gribbin, 1993).

Einstein's relativity theory is based on a smooth and continuous spacetime. This is ideal when considering large scale structures of the universe and their evolution. However, on a small scale, we know that quantum mechanics is the correct theory. Quantum mechanics is the study of the physics of the very small. The physical phenomena on the very small scale (things less than an atom in size) behave in radically different ways to the physics of the larger scale, that we experience every day. However, there are quantum mechanical effects that can have profound impact on the macroscopic world. We touched on this previously when introducing the concept of cosmological inflation, which is explained on a quantum mechanical level. One can imagine the need to unify Einstein's theory of gravity with that of quantum mechanics, see

(Hossenfelder, 2013). Scientists have long pursued a theory of 'quantum gravity' but, unfortunately, there does not appear to be a successful theory to date. Some candidates are so-called string theory (Wood et al., 2022), quantum-loop gravity theory (Sutter, 2022) and twistor theory (Musser, 2010).

There are many attempts at constructing other theories of the universe in an effort to explain a list of unexplained phenomena. A class of such theories are called modified gravity. The general idea is based on the assumption that either Newton's laws of motion or Einstein's theory of General Relativity is deficient in some way(s). Among these are the so-called Modified Newtonian Theory (MOND). It was first introduced by Israeli physicist professor Mordehai Milgrom. The idea here is based on a new effective gravitational force law that relates the true acceleration of an object. It applies to objects with extremely low acceleration like stars, suns and planets. It is based on the modification of Newtonian mechanics, that is, on Newton's laws. However, Modified Newtonian Theory is not a full theory and is just a modification of some of the equations that constitute classical mechanics. In this approach, gravitational force varies linearly with the separation distant between two masses. The only current issue with Modified Newtonian Theory (MOND) is that, perhaps, it has not attracted the attention of many researchers. This is despite the enormous following commanded by the issues of dark matter and dark energy, see (Bowler, 2022).

An example of a modified theory that attempts to generalize Einstein's theory of General Relativity is so-called f(R) gravity, see ($f(R)$ gravity, 2022). The f(R) means there is a mathematical function, denoted by f, of the Ricci scalar, denoted by R. Roughly speaking, the Ricci scalar represents the curvature of spacetime. So, f(R) gravity represents as many members of a family of solutions as there are suitable functions, f. For example, the choice of function $f(R) = R$ is just General Relativity. The appeal in this approach, first introduced by German-Australian physicist Hans Buchdahl, is the arbitrariness of the function, f, permits great freedom to explore and possibly solve known issues that cannot be addressed using the standard theory. Of course, the more complicated the choices for the arbitrary function, f, will result in more complicated field equations. These are the equations one must solve in order to present solutions to a given problem. It is important to realize, however, that with this newborn freedom comes the reality that not all terms in $f(\mathrm{R})$ will pass observational tests nor satisfy physical constraints, see (de Felice et al., 2010).

The $f(\mathrm{R})$ modified gravity approach has attracted a large following and has enjoyed some success. Most notably, Alexei Starobinsky a Russian physicist, was able to resolve issues associated with inflation by introducing a corrective term within $f(\mathrm{R})$ gravity, namely

$$f(R) = R + \frac{R^2}{6M^2}$$

where M represents mass in units of kilograms. Recall, the first term in this expression, which is just the Ricci scalar R, represents standard General Relativity. So, the second term, which is quadratic in the Ricci scalar, R, represents the correction term. This entire expression represents so-called Starobinsky inflation which has been successful in passing all recent observational tests, for example see (Pallis, 2017).

Roger Penrose has also constructed an alternative theory of the universe that requires the existence of a previous universe, so-called Conformal Cyclic Cosmology, see (Penrose, 2011). The general idea is that a pre-Big Bang universe existed that kept expanding until it turned into the Big Bang event. His method extends Einstein's theory of General Relativity to include conformal mappings. This extension is needed because there is a transition from the previous universe to the current universe that cannot be accomplished in General Relativity alone. Basically, the infinite conformal time of the previous universe is mapped to the Big Bang event of our universe. The 'conformal time' is used as a mathematical tool and is assumed to apply in reality. However, the vast majority of researchers seem to object to this proposal on a number of theoretical grounds and on a lack of experimental evidence.

First, a requirement of the conformal mapping is that no mass can be left in the previous universe (before making the transition to our universe). The major problem here is that although some massive fundamental particles do have finite lifetimes, others do not. For example, protons do in fact have finite lifetimes, but electrons do not. So, one must resign to the hope that somehow particles like electrons can undergo total annihilation with their anti-particles. Or somehow an anti-Higgs field is generated that removes mass from the universe entirely. This process would be completely opposite to the Higgs field previously discussed during the early universe. The other issue stems from the fact that the theory assumes information loss in Black Holes (Black hole information paradox, 2022). It was Stephen Hawking, (see Illustration 4), who corrected himself on this point, conceding the Thorne–Hawking–Preskill bet in 2004, see (Thorne–Hawking–Preskill bet, 2022). Many scientists reject Penrose's proposal on this point alone.

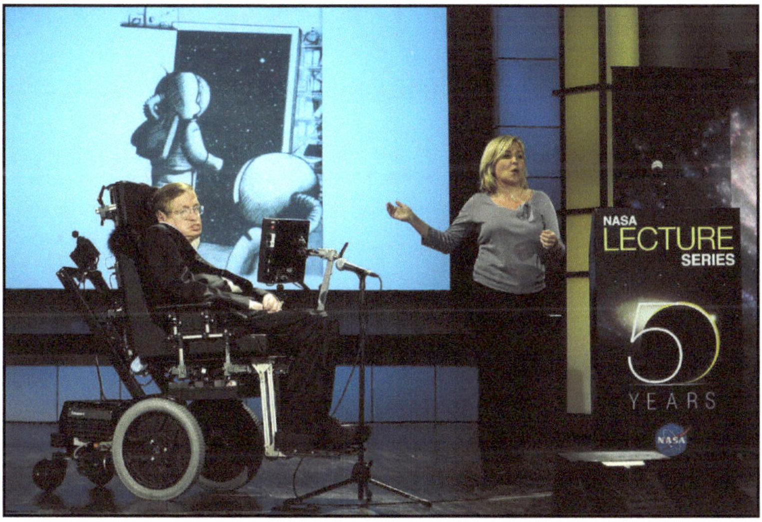

Illustration 4: Cosmologist Stephen Hawking. Image Credit: NASA/Paul Alers. https://www.nasa.gov/image-feature/stephen-hawking-and-our-changing-ideas-of-the-universe

The final objection is the lack of confirmation of statistical results in the search for concentric rings in the cosmic microwave background (CMB). Such rings on the cosmic microwave background are expected, as only bosons may transition from the previous universe to our universe, not fermions. Recall, from the chapter on particle physics, bosons and fermions are the two broad classifications of fundamental particles. They are classified by the particle spin, whether it is half-integral spin (fermions) or integral spin (bosons). Unfortunately, the evidence that such rings are statistically significant is lacking, see for example (DeAbreu et al., 2015). This basically means that one can find almost any shape in the CMB data, but is that just by chance or a result of something deeper? If it is due to something deeper, then one should expect it to be statistically significant: perhaps due to a mechanism that is responsible which leaves its signature on the CMB, like the before the Big Bang scenario that is being proposed.

Even if you restrict all alternatives to General Relativity to include only well-known and established members, the list would still be extensive. It may be broken down into the general mathematical technique used to construct the required field equations or appropriate laws. For example, General Relativity itself belongs to a general class of theories called tensor theories. This is because the theory relies on the use of tensors, complicated mathematical objects that require associated techniques for manipulation, to construct the field equations. Another member of this class is f(R) gravity that we considered above. The list of classes for alternatives theories, by no means exhaustive, includes scalar field theories, quasi-linear theories, scalar-tensor theories, vector-tensor theories, bimetric theories and non-metric theories, see (Moskvitch, 2018).

One could conclude, from surveying various efforts in producing a complete theory of our universe (which must include gravity, of course), that the standard model (LCDM) is on a sure footing. To date, nothing can claim to have completely superseded it, and all attempts to do so have either failed or can only partially contribute in some areas. It has been long said that String Theory would take that crown but research efforts in its development, as measured by publication rates, seem to have dramatically reduced over the last few decades or so. This is despite that fact that there are glaringly significant fundamental questions that have yet to be answered, see (Smolin, 2013).

ASPECTS OF COSMOLOGY

CHAPTER 6
THE BIG BANG

There was a 'point' in space and time, where and when, our universe began. This is a difficult concept to understand because there is no reference before this event. The universe simply came into existence in a very special way for no explainable reason. This is the ultimate mystery of physics, and more broadly science in general. Many scientists have conjectured on the origins of the universe. A popular view is based on the quantum mechanical process of spontaneous existence or spontaneous creation. Some even believe that the universe created itself.

> *"Because there is a law such as gravity, the universe can and will create itself from nothing,... Spontaneous creation is the reason there is something rather than nothing, why the universe exists, why we exist."*
>
> — *Professor Stephen Hawking,* The Grand Design

The point we referred to above is formally called a singularity. The Big Bang singularity is estimated to have occurred over 13.7 billion years ago, see (Big Bang, 2022) and (Figure 12,). Think of it as the biggest explosion ever to have taken place at, literally, a point in 'nothing'. We don't actually know what it started in; our best guess is absolutely nothing! Although there are some scientists that have proposed, self-confessed, 'crazy theories' about this issue, none are accepted by the mainstream. More than that, we certainly don't know why it happened. The space and time we live in today was born out of that singularity, see (Gravitational singularity, 2022). Space grew out of that point in nothingness and has continued to expand. It's very important to understand that our universe is expanding in a background of, well, nothing that we can possibly understand, or so it would seem, see (Choi et al., 2022). We are assuming, of course, there is indeed a background. Think of a balloon expanding due to warm breath being channeled into it. The surface area of the balloon is expanding. If two-dimensional biological entities lived on its surface, they would understand their balloon universe was expanding, but not that it was expanding in the background of our Earth's atmosphere.

Figure 12: Artist impression of the Big Bang. Credit: Wikimedia.org by Pablo Carlos Budassi. https://commons.wikimedia.org/wiki/File:Bigbang.png

What happened at the Big Bang is another matter. The theory that proposed this incredible phenomenon is none other than Einstein's theory of General Relativity, introduced previously. His theory naturally predicts a Big Bang scenario followed by an expanding universe, see (Howell et al., 2022). There was no need for extra information to be added externally, and so it triumphed against the strongly held belief of a steady-state universe. The steady-state theory of our universe assumed that the density of matter was fixed because the expanding universe was balanced by continuous creation of matter. Put another way, the observable universe is indistinguishable no matter where or when you make measurements. The beginning of the downfall of this picture was the discovery of the cosmological microwave background radiation. (We will investigate this in greater detail later.) The fact that the steady-state model does not predict a Big Bang scenario is cited as why most modern scientists no longer believe its validity.

Let us consider this event and its implications more closely. As a thought experiment, imagine lighting a firecracker as a way of celebration at a New Year's party, (see Illustration 5.). From the point of view of an observer far enough away, a small firecracker would appear as a point in three-dimensional space. Then you light it and throw it into the air. It explodes with expanding flashes of light, quickly filling the space around it in a spherical shape. The explosion would appear uniform and smooth in some sense. That is, all the bright flashes would be uniformly distributed in a spherical shape. If this shape was big enough (imagine a huge firecracker), the flashes would look evenly distributed no matter which direction you looked. The Big Bang event of our universe can be thought of in the same way.

Illustration 5: Fireworks. Notice the spherical shape of the explosion with the hot core of the firecracker at the center. https://commons.wikimedia.org/wiki/File:Summerfest_2008_fireworks_7077.jpg

The Big Bang singularity can, perhaps, be better understood if you 'run the film backwards'.

Suppose we explain what we mean with a mathematical game. We let the current state of the universe be represented by a fraction, say 1/4, which is just a number. The denominator of this fraction decreases as we start playing the film, of the expanding universe, backwards. Imagine, each second the film plays backwards results in halving the value of the denominator. After the first second, the fraction becomes 1/2. This is a larger number than 1/4. After the next second, we have 1/1 (=1), which is larger than 1/2 or 1/4. Eventually, after a long enough period, as we watch the Big Bang singularity in reverse, the fraction will become a larger and larger number. This pattern continues, from 1/(1/2), 1/(1/4), 1/(1/8), ..., 1/(1/1024) and so on, until the film stops, (see, Illustration 6.). The value of the fraction can be described as 'blowing up' because its denominator is approaching zero. Of course, this just represents the basic idea of a Big Bang singularity but should give you some idea on what it represents, see (Sutter, 2021).

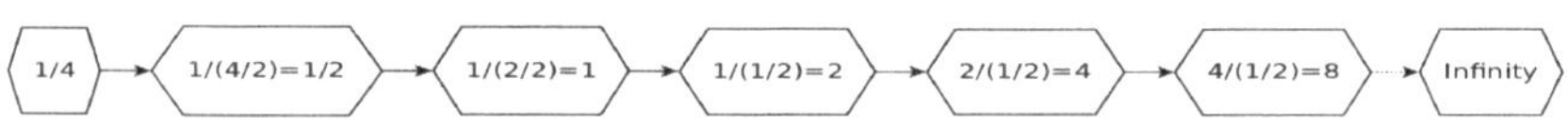

Illustration 6: Starting with an arbitrarily chosen fraction to represent a given state of the universe at a given time, say 1/4. The denominator is then divided by a factor of 2 each frame of a film of the universe's evolution plays backwards. This helps illustrate the nature of the Big Bang singularity.

We introduced the concept of Entropy earlier that plays a significant, and not fully understood, role at the Big Bang. Some scientists believe unveiling this mystery holds the key to a full and deep understanding of our universe, see (Siegel, 2017). One of the foundational laws of entropy is that it always increases in time. This suggests it was at a relatively low value a brief time after the Big Bang. This epoch was in a uniform and highly dense state. Given the violent nature of the Big Bang, you would be surprised to learn that the early universe was near thermodynamic equilibrium. This leads to a serious problem, since it is hard to achieve such a homogeneous (or uniform) state if there are regions of space that are not causally connected. This problem has been named the Horizon problem after it was pointed out by the physicist Wolfgang Rindler in the mid-1950s, see (Horizon problem, 2022). This problem will be discussed further at a later stage.

This means that the Big Bang event had to be, almost, impossibly special. There must be a mechanism that sets the same initial conditions everywhere. This mechanism is thought to be cosmological inflation, introduced previously. You may think that the entropy of this state should be high, and normally you would be correct. However, it is in fact low since gravitational attraction plays a major role. You will remember from our previous discussions that this is not the case for systems where gravitational attraction plays little or no role. Thermodynamic equilibrium is associated with a state of maximum entropy in systems where gravitational interactions are omitted. But the Big Bang event is even more special because, due to its compact size and future expansion, the entropy must have been lower. Remember, as gas expands, the entropy must increase. Put another way, the early universe should be considered highly ordered and so has low entropy.

Cosmological inflation has been celebrated by most scientists, but not everyone has a blind enthusiasm when it comes to understanding the details. One of the world's most leading mathematical and theoretical physicists, and the first physicist to be awarded a Fields Medal, Professor Edward Witten, believes inflation was at work in the early universe. The exact reasons behind this mechanism are still unknown because there is simply no way to currently perform experiments under the same conditions. The energies that were at play in the early universe simply dwarf anything that modern science can possibly achieve. Just as important is the need for an understanding of how gravitational physics and quantum mechanics behaved under such conditions. This would, of course, require a full theory of quantum gravity, see (Becker, 2022). As we touched on previously, such a theory is currently not available. Inflation is a simple idea, and Edward Witten believes it has no viable challengers.

> *"As for inflation, scientists believe it occurred in the early universe at a temperature far above the energy of particle accelerators and relatively close to the energy at which quantum gravity becomes important. We do not yet have a convincing, detailed model of how and why inflation transpired because the models of particle physics that we have are not adequate at the huge energies of inflation. Understanding inflation requires a much better knowledge of particle physics than we have now, and possibly a full knowledge of string theory and quantum gravity."*
>
> — *Professor Edward Witten*, (June 2002)
> "Universe on a String". Astronomy Magazine: 42–47.

We will revisit cosmological inflation later, and dive into some more details associated with its quantum mechanical nature, also see (Inflation (cosmology), 2022).

CHAPTER 7
WHAT IS A BLACK HOLE

The theoretical conception of a Black Hole has a long and fascinating history, dating back to Einstein himself. Black Holes are only predicted from Einstein's theory of General Relativity, see (Tillman et al., 2022). Newton had thought a dense enough celestial object could not possibly exist to form something like a Black Hole. This fact comes directly from Newton's formulae of the law of gravitation and classical mechanics. The equal and opposite pressure formed from a collapsing celestial body, like a sun at the end of its life cycle, will find an equilibrium state. This means it will not keep collapsing forever. In contrast, such a collapse is predicted to continue to create what is called a singularity in General Relativity. Remarkably, John Michell and Pierre-Simon Laplace, centuries before Einstein, dreamed of fictitious planets where the gravitational fields were so strong light could not escape, see (Black hole, 2022).

One way we can think of a Black Hole is to imagine the Big Bang in reverse. The Big Bang is an explosion, expanding the fabric of space and time, whereas a Black Hole is formed from an implosion, contracting the fabric of space and time. It was the famous physicist Stephen Hawking that first constructed this picture whilst working with Nobel Prize laureate Roger Penrose. It turns out that similar mathematical tools can be used in either of these physically distinct phenomena.

A Black Hole is basically created from an exhausted star of some kind, like our own Sun, see (Tillman et al., 2022). The birthing process must involve some kind of celestial body that is in the process of radiating out energy, typically something that occurs in suns, (see, Illustration 7.). When the fuel that is responsible for this radiation runs out, given the weight of the star, it starts to collapse. Most importantly, if the mass of the star is sufficiently large, it will continue to collapse until it forms a Black Hole. There are some very subtle details, on an atomic scale, that make this process work. However, we shall glean over this point to avoid unnecessary complications in our discussion.

A very important point, that needs highlighting, is associated with the degree to which the collapse takes place. That is, does the collapsing star continue to collapse forever? Well, mathematically, according to the equations of the theory of General Relativity, the answer is "Yes"! This is exactly what we mean by the term singularity. The idea behind the concept of a singularity may be understood by considering a fraction with an ever-decreasing denominator. Here is an illustration in mathematical language: consider the fraction 1/16, it's just some number, right. Now decrease its denominator, namely the number 16, by half. We now have the fraction 1/8. What we notice is that 1/8 is larger than 1/16. Now continue this process by halving 8, this leads to the fraction ¼, which is larger than 1/8. So, as you can quickly see, this process will produce larger and larger numbers as it continues. In fact, as the denominator approaches zero, the value of the fraction will approach infinity, see (Sutter, 2021). Scientists describe this colloquially as 'Blow-Up', or more scientifically, a mathematical singularity, see (Singularity (mathematics), 2022).

Illustration 7: Artist's illustration of black hole. Artwork Credit: NASA, and M. Weiss (Chandra X -ray Center). https://www.nasa.gov/content/black-hole-friday

The fact that the theory of General Relativity exhibits singularities has led many scientists to claim a break-down of the theory at such places, see (Sutter, 2022b). At the center of every Black Hole that has been born out of a collapsing star (of sufficient mass) is a singularity. A place where the theory fails! This was such a profound consequence for Einstein himself that he simply did not believe it to be true.

There is yet another important point to be made on this topic. The so-called singularity of a Black Hole is not 'naked', see (Sutter, 2020). It is, in fact, shielded by a spherical surface evenly about the singularity called an event horizon. You may think of this as the 'point of no return'. Once crossed, there is no turning back. Whatever enters will never come out (at least, not in the exact same form for Earthly bodies, such as Humans!). The event horizon plays an essential role in Black Hole physics and is intermittently connected with the concept of Entropy, introduced previously. In fact, the event horizon must always cover the singularity, according to the so-called cosmic censorship hypothesis proposed by Roger Penrose, see (Penrose, 1969). This basically states that no 'naked' singularities exist in the universe, see (Naked singularity, 2022).

How Black Holes are formed through collapsing stars is a fascinating process and of great importance, see (Emeka-Okafor, 2021). Let us consider the entire process using our Sun as an example. Our Sun is a heat engine of sorts, its fuel being Hydrogen and other gases. After a given time, these gases are expected to deplete, just like the fuel in your car. Unlike your car, however, there is no cosmic fuel station to replenish the Sun's gases. At that point in time the Sun will start burning its core, you can think of this as the actual engine. If this happened to your car, then that would be bad news, indeed. This is also the case for our Sun, it would work overtime and expand in size, devouring all planets in its path, including the Earth. The Sun would be going through, what could only be described as, its death throes before collapsing in on itself. If the collapse continues and reaches, and eventually crosses, the event horizon, then a Black Hole will form. Interestingly, its gravitational field will remain the same, since the mass of the now dead Sun has not changed by any significant amount. Of course, our Sun could not collapse to form a Black Hole, since it has insufficient mass, see (Fryer, 2003).

In general, any stellar object like a star can collapse to form a Black Hole. The reason the collapse takes place, at all, is due to the overwhelming mass of the collapsing star. One could imagine the absence of such a collapse if the total mass of the star is insufficient. The idea would be that there is not enough gravitational pull, due to the reduced mass of the star, to overwhelm the chemical bonds of the matter itself. While thinking along these lines would be correct, as it turns out, there is a mass called the critical mass. The absolute minimum mass that the star must have in order to be a candidate to create a Black Hole. This minimum mass is called the Tolman–Oppenheimer–Volkoff limit and is currently estimated to be approximately just over 2 solar masses. The solar mass is a standard unit of measuring cosmological mass and is approximately 2×10^{30} kilograms (that is 2 multiplied by the number 1 with 30 zeros, see the appendix). It is approximately equal to the mass of our Sun. This means 2 solar masses is equivalent to two of our Suns in mass, see (Tolman–Oppenheimer–Volkoff limit, 2022).

You may well ask, what happens if I were to fall into a Black Hole, see (Betz, 2020). What would I see? Where would I go? Well, the answer leads to some very bad news. First and foremost, you would be destroyed as your spacecraft nears the event horizon. There are forces, called 'tidal forces', that will treat you like you're nothing more than spaghetti. Your feet would get pulled towards the event horizon at a greater rate than your head, thereby stretching your body, (see, Figure 13.). This will happen while the sides of your body will get squeezed with ever-increasing force. When you eventually cross the event horizon, the mass associated with what was once your body would undergo a remarkable experience. As soon as it crosses the event horizon, which is effectively invisible, the roles of space and time will get interchanged. This means the mass, that was your body, would keep falling towards the singularity because time moves forward.

A bizarre phenomenon happens as your spacecraft is hurtling towards the event horizon of a Black Hole. Suppose you had a friend who was observing your journey. In order to ensure the safety of your friend, you instruct them to stay a very far distance from your location. They then record your journey but find something very odd. They report that you never crossed the event horizon. You simply approached it at a slower and slower rate until your spacecraft appeared motionless at the event horizon. How could this be possible? This is yet another fascinating property of Einstein's theory of relativity. The journey takes place with two completely different observations, from each observer, that are both correct. The event horizon is both an invisible shell around a singularity and an impassable 'brick wall'. It just depends on who you ask, see (Betz, 2020).

Figure 13: Figure showing tidal forces acting on person falling towards the singularity of a Black Hole. They would feel a stretching force between their head and feet, and a crushing force on their sides. This happens as the Black Hole tries to suck them into its singularity.

Yet another important question to ask is: if no light can escape from a Black Holes, then how do you know they are there or even existed? A black Hole would simply appear as a dark dot in an otherwise starry night. It would be like trying to find a needle in a haystack. Well, it turns out it's not that hard to figure out what to look for, and scientists achieved this in c.1979. The theoretical idea is call gravitational lensing and, as the name suggests, works by assuming Black Holes warp spacetime. This can be thought of in much the same way as an optical lens, say in a pair of spectacles, warps or distorts visible light. One would then scan the observable universe in search of planets and stars that appear displaced from known positions. That is, planets in the background of a Black Hole, with respect to an observer here on Earth, will see a spacetime distortion effectively displacing its true location. This is an important and fundamental result, that not only verifies Einstein's theory, but also plays an important part in the missing mass and energy of the universe.

Up until the mid-1970s, Black Holes were thought to be completely black, literally nothing could escape their gravitational attraction. Even light was unable to free itself. This is remarkable considering visible light is made from particles called photons. These particles have no mass, which is the reason they can travel at the speed of light. Unexpectedly, a brilliant physicist by the name of Stephen Hawking turned the conclusion completely around. He managed to unify three areas of physics at the event horizon of a Black Hole. Namely, Einstein's General Theory of Relativity, Thermodynamics and Quantum Mechanics. In a stroke of genius, he showed that Black Holes do in fact emit thermal radiation, see (Koberlein, 2021).

So-called Hawking Radiation is expected to slowly evaporate Black Holes, albeit very slowly. We know that Black Holes can gain mass by absorbing other objects through gravitational attraction. However, as the universe expands, one could imagine that this process of increasing mass would run down. If this was the case, then total evaporation would only be a matter of time, (see, Illustration 8.). That said, all but the smallest Black Holes would take an extraordinary length of time to vanish. This is because the temperature at which a Black Hole radiates is inversely related to their mass. In other words, the bigger the mass, the smaller the temperature.

Stephen Hawking was, in part, inspired by the work of the brilliant physicist Jacob Bekenstein, see (Bekenstein, 2019). Bekenstein thought about the consequences of throwing a cup of coffee into a Black Hole. He conjectured, a few years before Hawking's discovery, that the black holes should have entropy. Otherwise, entropy is lost from the universe through Black Holes. This is related to the so-called Black Hole information paradox, which has long raised strong concern within the scientific community. Simply put, this means that information from our universe could potentially be deleted, lost forever, in a Black Hole. However, although Hawking initially thought this process was indeed possible, he later changed his opinion on the topic and agreed with Bekenstein. Hawking Radiation is a profound result and has far-reaching consequences that we will consider later.

Illustration 8: Radiation from supermassive black holes at the cores of galaxies. Image credit: NASA/JPL-Caltech. https://www.nasa.gov/jpl/nustar/pia18919

We previously discussed that the concept of Entropy was derived from a statistical framework. In a cosmological setting, the Entropy associated with a Black Hole is encoded into its event horizon. However, a Black Hole's event horizon is a surface (think of it as a spherical empty shell, like a basketball with infinitesimally thin skin) and therefore is two-dimensional. You may recall that Entropy is a so-called extensive quantity and therefore associated with volume. In other words, it is three-dimensional. How can this be true? Well, Bekenstein helped resolved this issue by proposing a bound on the maximum Entropy that can be contained within a Black Hole which has a finite amount of energy, (Bekenstein, 2019). Remarkably, this bound does not involve the critical role of gravity itself. This idea inspired the holographic principle, which was first submitted by Dutch theoretical physicist, and 1999 Nobel Laureate in Physics, Gerardus 't Hooft. This is the idea of projecting the Entropy, in terms of a volume which is three-dimensional, onto a two-dimensional surface. The two-dimensional surface, that is the event horizon, reflects all the information from the three-dimensions. This is basically what happens with a holographic, optical, image that gives the appearance of being three-dimensional.

Scientists theorize that every galaxy has a supermassive Black Hole at its center, (see, Illustration 9.). Typically, these Black Holes are colossal in mass, some well over 1 million solar masses. Even more remarkably, some galaxies have two supermassive Black Holes. This is called a binary supermassive Black Hole system. Given the gravitational attraction of both Black Holes, they are locked in an in spiral motion. One would expect such a system to eventually lead to merger. Recently, galaxy NGC 7727 has been observed to display extended spiral arms, which is evidence of a merger in process, see (de Lazaro, 2021). The two Black Holes of this galaxy are 154 million and 6.3 million solar masses, respectively. They are only 1,600 light-years apart. Unnervingly, NGC 7727 is only 89 million light-years away from our own galaxy. The energy associated with the merger takes a backseat only to the Big Bang itself.

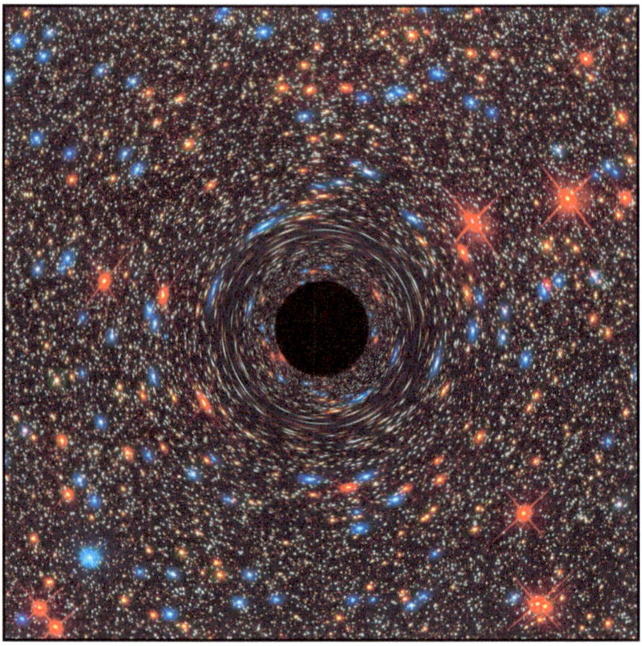

Illustration 9: This computer-simulated image shows a supermassive black hole at the core of a galaxy. The black region in the center represents the black hole's event horizon, where no light can escape the massive object's gravitational grip. The black hole's powerful gravity distorts space around it like a funhouse mirror. Light from background stars is stretched and smeared as the stars skim by the black hole. Image Credit: NASA, ESA, and D. Coe, J. Anderson, and R. van der Marel (STScI). https://www.nasa.gov/image-feature/computer-simulated-image-of-a-supermassive-black-hole

If the universe keeps expanding, then Black Holes will eventually absorb all objects and even merge with each other before evaporating themselves. They will eventually all vanish with a 'pop'. However, what if the universe does not keep expanding? What if it starts contracting after a period of expansion. We will consider the ramifications of this possibility later in greater detail.

Einstein's theory of General Relativity presents field equations that are stupendously difficult, if not virtually impossible, to solve in general. In response, mathematicians and physicists developed analytical tools in order to sufficiently reduce the complexity of these equations. This enabled them to then solve the field equations exactly. By exactly, we mean without approximation, perturbation or using numerical techniques. One of these tools is based on the assumption of, so-called, local symmetry. The concept of symmetry in this regard simply means things don't look different if you move along a certain direction. One of the simplest symmetries that is directly relevant to cosmology is spherical symmetry. This simply means the symmetry associated with a 'ball'. Think about being on the surface of a smooth ball. No matter which way you move, from one pole to the other, or along the equator, or another combination, what you will see does not change.

The assumption of spherical symmetry reduces the complexity of the field equations. But, just as importantly, it is also relevant in cosmology because it is a good approximation to the general shape of celestial objects like stars, planets, and especially Black Holes. German physicist Karl Schwarzschild, in the early 20th century, considered this when attempting to solve the field equations of Einstein's theory of General Relativity. He also assumed that the solution did not change as a function of time. This simply meant it was static in its evolution. When Schwarzschild put these assumptions, and a few others, into the field equations, he managed to solve them exactly. This was an incredible achievement as it would be interpreted as an exact solution for a static spherical systematic Black Hole. Einstein himself resigned to the fact that trying to solve the field equation of General Relativity was a fruitless pursuit. He was subsequently surprised to learn of Schwarzschild's result, see (Bartusiak, 2016).

> *"I have read your paper with the utmost interest. I had not expected that one could formulate the exact solution of the problem in such a simple way. I liked very much your mathematical treatment of the subject. Next Thursday I shall present the work to the Academy with a few words of explanation."*
>
> — Professor Albert Einstein.
> (Eisenstaedt, 1989).

What enables the Schwarzschild solution to be a model for a static spherically systematic Black Hole is the so-called Schwarzschild radius. This radius is

$$R_s = \frac{2GM}{c^2}$$

where G is the gravitational constant, M represents the mass of the ball and c is the speed of light. If the mass of the ball is sufficiently large, then a Black Hole solution is generated. It is called the Schwarzschild Black Hole. The importance of the Schwarzschild exact solution cannot be emphasized enough, as it has led to incredible insights into the workings of Black Holes.

This is not the end of the story as far as these types of solutions to Einstein's theory of General Relativity is concerned. The Schwarzschild solution, despite being an incredible achievement, lacks one vital feature, it does not rotate, (see, Illustration 10.). A Black Hole is formed by an extremely violent process during the collapse of a star. One can imagine that stars are often in a state of rotation, just like the Earth. It would be almost unimaginable if somehow this highly dynamic and chaotic process would lead to a non-rotating Black Hole. Intuitively, one could see that the Black Hole would have to be rotating to some degree.

Illustration 10: Supermassive black hole at the center of spiral galaxy M81. mage Credit: X-ray: NASA/CXC/Wisconsin/D.Pooley & CfA/A.Zezas; Optical: NASA/ESA/CfA/A.Zezas; UV: NASA/JPL-Caltech/CfA/J.Huchra et al.; IR: NASA/JPL-Caltech/CfA. https://www.nasa.gov/content/black-holes-have-simple-feeding-habits

One of the most remarkable achievements in General Relativity, and certainly in mathematical physics, is the discovery of the Kerr solution. Roy Kerr, in 1963, found an incredible solution of the field equations. This solution was similar in some aspects to the Schwarzschild solution but with one extra important feature, it was rotating! If the rotation parameter, which indicates the speed of the angular momentum, of the Kerr solution is set to zero, then the Schwarzschild solution is recovered. Put another way, the Kerr solution could be thought of as the Schwarzschild solution that has started to rotate. So, in summary, the Kerr solution can represent a rotating Black Hole of a given mass. At that time, some scientists strongly believed that Black Holes modeled by the Kerr solution were expected to be found in our universe, and they eventually were, see (Shipman, 1975).

Shortly after Kerr's celebrated discovery, scientists found that an overall charge could be added to the Kerr solution. Ezra T. Newman, and his research team, achieved this by solving Maxwell's equations of electromagnetism, for the Kerr solution, using a mathematical trick. This solution is called a charged Kerr solution, or more commonly a Kerr-Newman solution. So, a Black Hole could be modeled by a Kerr-Newman solution of the field equations with three parameters. The three parameters represented the Black Hole's mass, rotation, and charge.

Researchers quickly discovered the importance of the three parameters of mass, rotation, and charge. It turned out that these three parameters are all that is needed to fully describe the physical characteristics of a Black Hole. Scientists referred to these incredible results as the 'No Hair' theorem, see (No-hair theorem, 2022). After the complex processes that are involved in the development of the Black Hole end, there is no need for any more information. Remarkably, there are no other features of a Black Hole that distinguish it in any way. This means that two Black Holes with the exact same values for their mass, rotation, and charge would be (at least theoretically) identical, see (Johnston, 2019). This is a remarkable instance where theory, born out of the thoughts and minds of humans, exactly predicts the physics of an object in our universe.

"*… the unique solution which the general theory of relativity [or General Relativity] provides for the description of all black holes that can occur in the astronomical universe by the gravitational collapse of stellar masses; and it is the only instance of a theory providing an exact description of a macroscopic object.*" – Subrahmanyan Chandrasekhar, The Mathematical Theory of Black Holes

The possible mass ranges of Black Holes vary substantially. There are no Black Holes with mass less than approximately two time the mass of our Sun, as previously discussed. Remember, Black Holes are formed through the process of core collapse, either directly or in some cases indirectly, see (Carroll et al., 2017, Chapter 15). Specifically, these are called stellar-mass Black Holes.

However, there are other types of Black Holes, much larger in mass, whose formation is not fully understood. Moreover, we will introduce yet another possible type called a Primordial Black Hole in the next topic where this lower bound on the mass no longer holds.

Stellar-mass Black Holes are considered the smallest in terms of their mass range. They can vary from about 3 to over 60 solar masses. The smallest of such Black Holes has been measured to have a mass equal to approximately 3.3 solar masses, see (Thompson et al., 2018). On the other end of the spectrum, the largest stellar-mass Black Hole was found due to the result of a merger of two smaller Black Holes. It was discovered during the first detection of the existence of gravitational waves, see (Abbott et al., 2016). The primary stellar-mass Black Hole was around 36 solar masses, while the secondary stellar mass Black Hole was approximately 29 solar masses. (Primary and secondary are simply referring to the largest and smallest of the two, respectively.) This produced a stellar-mass Black Hole of a combined mass of about 62 solar masses, after merger.

Intermediate-mass Black Holes may have a substantial mass range between 100 to 10,000 solar masses. As such, they are significantly more massive than stellar-mass Black Holes. They are thought to be formed in the presence of surrounding galaxies and globular clusters through coalescence or mergers between smaller, stellar-mass or, smaller intermediate-mass Black Holes, (see, Illustration 11.). The best evidence, so far, for such Black Holes is from the event that was observed by the LIGO and Virgo detectors of 2019, called GW190521 (GW=Gravitational Wave and 190521=date), see (GW190521). It was another gravitational wave signal resulting from the merger of two black holes. The primary Black Hole had a mass of around 85 solar masses, while the secondary Black Hole had a mass of approximately 66 solar masses. The resulting Black Hole has a combined mass of about 142 solar masses.

Supermassive Black Holes are thought to exist in the middle of all galaxies. They are colossal in mass, ranging from 100,000 to 1 billion solar masses. For example, in the case of our galaxy, called the Milky Way, there is a centralized Black Hole which has a mass in the order of 3.7 million solar masses. There is little understanding among scientists on how exactly these types of Black Holes form. Some front running ideas are; mergers between galaxies themselves and/or the merger between other smaller mass Black Holes, see (Carroll et al., 2017, Page 639).

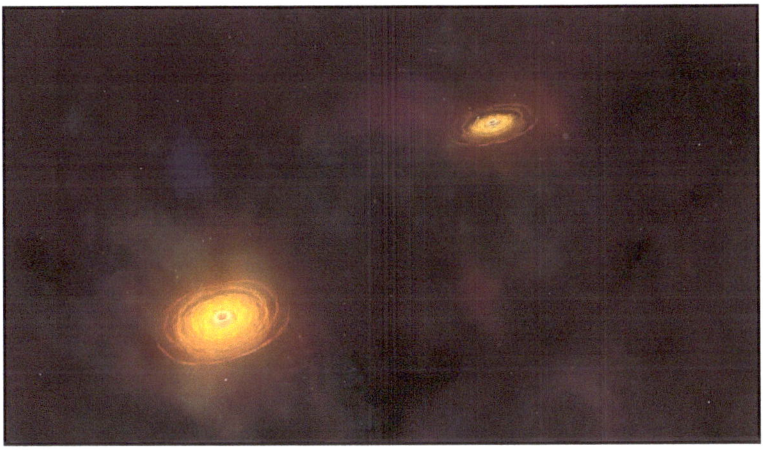

Illustration 11: Artist's illustration of a pair of supermassive black holes. Image credit: NASA/CXC/A.Hobart. https://www.nasa.gov/mission_pages/chandra/images/illustration-of-supermassive-black-hole-pair.html

CHAPTER 8
HAWKING RADIATION

Originally, when Black Holes were theoretically predicted, scientists thought they would simply last forever and never die out. This is because they absorb everything in the universe, including light itself. It was thought that nothing came out of a Black Hole by their very definition. This turned out not to be true. Stephen Hawking, in 1974, produced a remarkable and surprising result – Black Holes radiate. He managed to unify three areas of physics; General Relativity, Quantum Mechanics and Thermodynamics, to prove this result, see (Hawking radiation, 2022).

The radiation that takes place is considered minuscule compared to the total mass of a Black Hole, that are typically greater than 3 solar masses. This is because it involves the particles to be produced alongside their anti-particles. We earlier introduced the concept of anti-particles that were created at the Big Bang event, but this situation is rather different. It is the Black Hole itself that produces the particle and anti-particle pair. This means some mass of the Black Hole is being used in their production. Put another way, the Black Hole reduces in mass in order to produce the particles. This takes place outside the event horizon. At first this may seem strange because, even if a particle and anti-particle pair are produced on the exterior of the event horizon, they should annihilate each other. The associated energy of the production would simply be absorbed by the Black Hole again and no mass would be lost. This is in part true, however, there is another possibility that must be considered. One of the particles in the pair could first fall into the Black Hole. The other particle in the pair would then be jettison away from the event horizon and effectively escape the attraction of the Black Hole, see (Carroll et al., 2017, Page 644).

One can see, given the slow rate of this process, the radiation would result in a minuscule reduction of the mass of the Black Hole. This can be seen by the following estimate of the evaporation time,

$$t_R \approx 2.1 \times 10^{67} \left(\frac{M}{M_o}\right)^3 years,$$

where M is the mass of the Black Hole and M_0 represents a solar mass, see (Carroll et al., 2017, Page 136). The age of our universe is estimated to be 13.8 x 10^9 years. Assuming a Black Hole evaporated in this time, yields its required mass to be,

$$M \approx 8.69 \times 10^{-20} M_o \approx 1.73 \times 10^{11} kilograms.$$

This represents a tiny Black Hole which immediately rules out stellar-mass and larger types, see (Carroll et al., 2017, Pages 644-46).

One might conclude that the detection of Black Holes through Hawking radiation is impossible. Even a very small (smaller than a stellar mass) Black Hole of one solar mass would totally evaporate in an estimated **[Jacob 035]** years. This is so far into the future it might as well be considered an infinite long time period. However, that is not the full story since one must also consider another type of Black Hole called a Primordial Black Hole, (see, Illustration 12.). They are not classified in the same way as regular Black Holes previously considered and possibly have a huge range in mass. We will consider them in the next topic.

Illustration 12: Artist's impression of a growing supermassive black hole located in the early universe. Image credit: NASA/CXC/A.Hobart. https://www.nasa.gov/centers/marshall/about/star/star110622.html

CHAPTER 9
WHAT IS A PRIMORDIAL BLACK HOLE

There are two types of possible ways Black Holes can form, which lead to vastly different characteristics. These differences are profoundly important, as we will see later. The first type of Black Hole is created as a result of a collapsing star, as we have discussed previously. For example, a star that has exhausted all its energy and collapsed in-on-itself. In the General Theory of Relativity, as we have discussed in detail previously, this is predicted to create a Black Hole. There are even larger Black Holes that play a central role in galaxies that are formed in a way not fully understood by scientists.

The other type of Black Holes, that could exist, are formed in the early universe, see (Pultarova, 2021). They would not be created as a result of collapsing stars but a coalescence of matter and radiation just after the Big Bang. This is possible if there were sufficiently dense regions in the early universe undergoing collapse due to gravity alone. These types of Black Holes are called Primordial Black Holes (see, Illustration 13). They were first proposed by Soviet theoretical physicist Yakov Borisovich Zel'dovich and Russian theoretical astrophysicist and cosmologist Igor Dmitriyevich Novikov in 1966. However, it was not until 1971 when Stephen Hawking studied their properties that they became more popular. He developed a theory behind their origins and suggested some fascinating ideas. We will explore these ideas in more detail soon.

Illustration 13: Artist's illustration of what could be a primordial black hole. Image credit: NASA/JPL-Caltech. https://www.nasa.gov/mission_pages/spitzer/multimedia/spitzer20100317a.html

They are hypothetical, since there has been no observational or empirical evidence that they exist. Remarkably, there is no minimum mass associated with the possibility of their formation, like the Tolman–Oppenheimer–Volkoff limit. Therefore, Primordial Black Holes may be tiny, and have low mass, or gigantic in comparison. Technically, they could be as tiny as 10^{-8} kilograms to thousands of solar masses. However, primordial black holes that were formed with a mass lower than 10^{11} kilograms would have totally evaporated by now due to a phenomenon called Hawking radiation, as previously discussed. You may recall that we preformed the calculation that proved this fact previously. This naturally leads to the idea of trying to detect Primordial Black Holes with mass below 10^{11} kilograms. What we mean by this is, if something emits radiation, then it is possible in theory to detect it as a relic. Unfortunately, there has been no luck to date, and they remain a hypothetical idea, see (Carroll et al., 2017, Page 639).

Primordial Black Holes are being formed in the radiation era of the early universe. This can occur if there is a sufficiently dense coalescence of matter subject to large gravitational fluctuations, as proposed for example in (García-

Bellido, 2019). Remember, at this early stage of the universe there are no stars around to collapse into traditional, stellar-mass, Black Holes. There are only high energy particles and radiation, the gravitational attraction of which would have to overwhelm all the other three forces at play. Remember, there are four fundamental forces of the physical universe, namely the strong nuclear force, weak nuclear force, electromagnetic force and gravitational force. It is important to understand that the type of forces we see, and experience today would have been radically different in the early universe. In some sense they would have been unified at the very early universe, so to suggest one force would overwhelm the others is somewhat meaningless. Nevertheless, it is conceivable that at some stage of the very early universe this unified picture would no longer hold. Indeed, gravity would be characteristically likely to start to form Primordial Black Holes, see (Carroll et al., 2017, Page 645).

It is thought that the least massive Primordial Black Hole is given by the Planck mass,

$$M_p = \sqrt{\frac{\hbar c}{G}}$$

where G is the gravitational constant, c is the speed of light and $\hbar$ is the (reduced) Planck constant. The values of these fundamental constants are

$$G \approx 6.674 \times 10^{-11} m^3 . kg^{-1} . s^{-2},$$

$$c \approx 300000 km . s^{-1},$$

$$\hbar \approx 6.582 \times 10^{-16} eV . s$$

where $\hbar = h/(2\pi)$ is the reduced version of the Planck constant, denoted by h. The approximate numerical value of the Planck mass is therefore found to be

$$M_p \approx 2.176 \times 10^{-8} kg$$

Remarkably, the Schwarzschild radius for this mass, that is [Jacob 045] is given by

$$R_s^{(Planck)} = \frac{2GM_p}{c^2} = 2\sqrt{\frac{\hbar G}{c^3}}$$

However, we recognize that the last term involves the Planck length,

$$l_p = \sqrt{\frac{\hbar G}{c^3}} \approx 1.616 \times 10^{-35} m.$$

So, this means the least massive Primordial Black Holes have a Schwarzschild radius proportional to the Planck length, that is $R_s^{(Planck)} \propto l_p$. . Note that in the above, it might be helpful to understand that a photon's energy is defined in terms of its frequency multiplied by a proportionality constant called the Planck's constant, see (Carroll et al., 2017).

The huge range in the possible masses a Primordial Black Hole can have is actually a great opportunity. According to Hawking radiation, all Black Holes must evaporate and, eventually, given a long enough time period, will vanish. When we say, 'long time', it must be understood that the time taken for a Black Hole of a mass greater than 10^{11} kilograms to totally evaporate is longer than the actual age of the universe. This process takes so long because the Hawking radiation is minuscule compared to the mass of the Black Hole.

Illustration 14: The black hole, Cygnus X-1. It is shown as 'feeding' on a nearby star as it pulls matter and energy into its event horizon. Credits: NASA/CXC/M.Weiss. https://www.nasa.gov/audience/forstudents/5-8/features/nasa-knows/what-is-a-black-hole-58.html

This means that Primordial Black Holes of mass less than 10^{11} kilograms would have all radiated their mass by now. At the extreme end, the smallest of such Black Holes is thought to be based on the Planck mass, denoted by M_p, as introduced above. This means a Planck mass Primordial Black Hole would evaporate by

$$t_R \approx 2.1 \times 10^{67} \left(\frac{M_p}{M_o}\right)^3 = 2.75 \times 10^{-47} \, years.$$

This translates to just 8.66×10^{-40} seconds. So, a Planck mass Primordial Black Hole evaporates within a phenomenally short period of time. The evaporation rate, due to Hawking radiation, increases as the mass of the Black Hole decreases. We say that the relationship of the radiation rate is inversely proportional to the mass, see (Carroll et al., 2017, Pages 644-6).

The final stages of the evaporation of a Primordial Black Hole (or any other type of Black Hole) are the final burst of Hawking radiation. This is the so-called runaway evaporation and would end with a violent explosion.

MODELS OF THE UNIVERSE

Chapter 10
Standard Model of the Universe

In the 1990s, scientists put together a model that adequately described the universe up to that time. It was called the standard model, (Shears, 2012), or more scientifically known as Lambda-Cold Dark Matter (LCDM), see (Lambda-CDM model, 2022). As the name suggests, it is a combination of the General Theory of Relativity, with the inclusion of a cosmological constant, and Cold Dark Matter. This model does not involve the standard model of particle physics, therefore there is no explanation of Dark Matter. Dark Energy, a yet unidentified form of energy, is simply the renaming of what was known as the cosmological constant problem, see (cosmological constant problem, 2022).

The standard model contains three basic building blocks. The first being standard matter of the observable universe; this is the everyday objects we interact with here on Earth and all the celestial objects we observe in the cosmos. The second is a cosmological constant associated with dark energy; this is harder to think about, but we now know its value is small and non-negative. You may think of this as a mathematical object that represents something, called Dark energy, that we don't fully understand. Finally, cold dark matter (CDM); a totally unknown form of matter that scientists believe is missing and accounted for in theories of gravity, see (Cold dark matter, 2022).

One of the foundational assumptions of the standard model is the so-called cosmological principle, see (Shears, 2012). That is, the spatial distribution of matter in the universe is assumed to be both homogeneous and isotropic over a large enough scale. This is a very complicated sounding foundational assumption, but when broken down it is easy to understand. We truly need to come to an understanding of this assumption to be able to think about the universe properly. First, consider the word homogeneous, this basically means it looks the same in the current epoch of the universe. That is, the matter (made up of fundamental particles) and radiation (stemming from electromagnetic fields) of the universe appears evenly distributed. On the other hand, Isotropic means basically the same thing, except with respect to your (the person making the observation) position. In other words, celestial objects having a physical property which has the same value when measured in different directions.

This model includes the cosmological constant, commonly denoted by (Greek) Lambda, which represents dark energy (or sometimes called vacuum energy). Recall, it was Einstein himself who introduced this quantity into his formulae to force his theory of gravitation, General Relativity, to adapt to the world view of a steady-state universe (the prevalent model at that time). The standard model also features cold dark matter as something called non-baryonic; cool, but dissipation-less, with collisions that occur only through gravity.

This is, in fact, a model of the Big Bang cosmology, since it is the leading theory that provides a good confirmation of some of the observed properties of our universe. It makes major predictions that have been verified, such as: the existence and structure of the Cosmic Microwave Background (CMB); the large-scale structure in the distribution of galaxies; the observed abundances of atoms like hydrogen, helium, and lithium; and the accelerated expansion of the universe. The standard model can be extended by adding concepts and properties that are missing, as discussed previously, and known to be required, for a complete theory of the universe. This includes so-called cosmological inflation, see (Siegel, 2019). We will explore cosmological inflation in more detail later given its importance.

Naturally, any other elements that are thought to be critical to our understanding of the universe can also be added. Of course, this detracts from the original elegant and powerful model based on Einstein's theory of General Relativity because there is no insight in the additions. For example, one could simply add cosmological inflation, but this does not help to explain why and how inflation works. This is not really all that satisfying from a predictability point of view because that is what makes the theory so powerful in the first place. In order to achieve a more complete understanding, one would have to basically switch to a model based on quantum gravity.

CHAPTER 11
EXPANDING AND ACCELERATING UNIVERSE

The universe is currently expanding and accelerating, see (Schmidt, 2012). These two events are astonishing, to not only think about, but to ask how they are known. Scientists discovered that, relative to the position of our galaxy, other galaxies were red-shifted in all directions. This means they are moving away from us and in all directions. The phenomena of red-shifted and blue-shifted light waves has a simple-to-understand analogy in sound waves. Think about an ambulance moving towards you with its siren on. As it comes towards you, the pitch of the siren sounds high, whereas it shifts to low when the ambulance moves away from your fixed position. So, blue-shifted light waves are like the ambulance coming towards you, whereas red-shifted light waves are like the ambulance moving away. Hence, when the universe is expanding we measure red shift in all directions, see (Expansion of the universe, 2022).

When something expands in our everyday world, we imagine that something expanding 'into' the world. The expansion of the universe is not like this at all. Rather, the fabric of space is expanding together with the mass and energy from the Big Bang. This is hard to contemplate because it simply doesn't take place in everyday life, but this is exactly what is happening in our universe, (see, Illustration 15.). To come to terms with this unusual concept, remember our analogy with a balloon being inflated, as previously discussed. Think about the surface of the balloon only and not the air it is occupying. Imagine our three-dimensional space has been compactified into the two-dimensions on the surface of the balloon. As a two-dimensional 'balloon being', your galaxy is in the shape of a disc on the surface and other galaxies are being stretched away from yours. The balloon is inflating as our universe is expanding.

The Hubble–Lemaître law shows that the universe is expanding, see (Elizalde, 2021). It was most notably highlighted by the American astronomer Edwin Hubble as he demonstrated the expansion of the universe. Hubble did this by realizing surrounding galaxies were all red-shifted and therefore are lengthening their light waves relative to the Earth. Hubble showed a relationship between the departing velocity of a galaxy and the proportional increase of its distance from the Earth. At least that is how history commonly remembers this incredible discovery. For this reason, sometimes this phenomenon is simply referred to as the Hubble Law. However, the Belgian Catholic priest, theoretical physicist, mathematician, astronomer, and professor of physics Georges Lemaître also made the same observation several years prior. Of course, this is not an uncommon pattern in physics when attributing credit for novel ideas, see (Hubble's law, 2022).

The fact that the rate of which our universe is expanding, and is also accelerating, was only recently discovered in 1998. This discovery implies that the cosmological constant of the standard model is small in value but non-negative in sign. In other words, it cannot be negative otherwise the expansion rate of the universe could not be accelerating. This discovery was not born out of theory, but through experiment. The standard model, for example, did not predict that the universe should be accelerating in its expansion. In fact, this came as a surprise to most scientists. This has significant implications for all theories of the universe. For a start, it automatically rules out any theory or model of the universe which assumes a negative value of the cosmological constant. For example, the anti-de Sitter spacetime has a negative cosmological constant which is interpreted as positive pressure and negative energy-density, see (Cosmological constant, 2022).

Illustration 15: The distant universe. Credit: ESA/Hubble/NASA. https://www.nasa.gov/image-feature/goddard/2017/hubble-sees-galaxy-cluster-warping-space-and-time

CHAPTER 12
PROBLEMS WITH THE STANDARD MODEL

There are numerous problems with the standard model in providing a vivid and detailed picture of the actual universe, (Gomez, 2018). One of the most obvious issues is the rapid expansion, called exponential expansion, of the fabric of space within a very short time period just after the Big Bang. This remarkable epoch in the evolution of the universe is called cosmological inflation, previously mentioned. We will dive into more detail on this topic later. The standard model simply does not predict inflation, and it certainly cannot explain the phenomena. We don't know why it occurred and this is basically a failure of the model itself.

The standard model (Big Bang cosmology model without inflation) cannot account for thermal equilibrium in the early universe simply due to gravitational expansion. There is simply insufficient time that is required for such equilibrium, that is, for things to settle down. In the early universe, it was not possible for the two distinct regions to communicate by light signal. Consequently, two regions of the observable universe cannot have achieved thermal equilibrium because they never had the chance to come into causal contact. So, how could the universe have become both isotropic and homogeneous on a suitable scale? This issue is called the horizon problem, see (horizon problem, 2022). Unfortunately, the standard model cannot offer an explanation to these phenomena that are based on fundamental physics.

The density of matter in the universe necessitates a flat universe. You may recall that we previously discussed the Minkowski spacetime (see Chapter 4. Theories of Gravity and the Universe, paragraph 5). This is exactly what we mean, but with the inclusion of a cosmological constant and matter content.

Technically, this is called the Friedmann–Lemaître–Robertson–Walker (FLRW) cosmology. The entire geometry of the universe is flat, but this cosmological model accommodates the non-flat hyperbolic or spherical geometries. It was long speculated in what type of geometry we live in, but it appears to be flat. This is supported by observations of the Cosmic Microwave Background (CMB) that have shown that the universe is flat to within an error of a few percent. A few percent errors is another way of saying it is very accurate. As a result, the curvature (of space itself) due to the expansion of the universe could not be much greater than the matter content. Put another way, the curvature is predominately due to the matter content of the universe. Extrapolated into the past, this presents what is called a 'fine-tuning' problem because the contribution of curvature to the universe must be exponentially small. This conundrum is known as the flatness problem. Although the standard model can accommodate the necessary geometry of the universe, as observed by experiment, it cannot explain the flatness problem, see (Flatness problem, 2022).

Furthermore, there is the issue with so-called magnetic-monopoles. This issue arose from the famous 1931 paper by P.A.M Dirac where he proved that the existence of a magnetic-monopole implies that electric charge in the universe must be quantized from a quantum mechanical perspective, see (Pinfold, 2010). This is now known as the Dirac quantization condition. It turns out that the electric charge is quantized, this was proved without the need of discovering the existence of any magnetic-monopoles. Theoretically, they should be everywhere in abundance, but they have never been found. A complete theory of the universe must be able to explain this conundrum. What exactly are they? A magnetic-monopole is like a magnetic but with either a North-pole or South-pole, only. Now, imagine you are holding a regular magnet. One end is the North-pole and the other the South-pole. You may think that a 'mono'-pole can be created by cutting the magnetic in half. This would be incorrect, you would simply have two smaller magnets, each with a North-pole and South-pole. In fact, you could repeat this cutting procedure without achieving the goal of making a monopole.

It would seem that magnetic-monopoles would have to be created at the Big Bang event. Magnetic monopoles are hypothesized to be abundantly produced at high temperature and they should have persisted to the present day. They should be the primary constituent particles of the universe.

However, magnetic-monopoles have never been detected nor observed. The standard model is totally mute on this important point, see (Magnetic monopole, 2022).

One of the most important phenomena that the standard model does not account for is inflation, as previously discussed. Inflation provides a dynamical mechanism, which is theorized to start at the quantum level, that drives the universe to a finely tuned special state. This leads to a universe identical to the one we observe at present. By 'finely tuned' we mean that if some aspect of the initial conditions of the inflation process were to be changed in the early evolution of the universe, it would change the outcome. That is, our universe may well look entirely different to what is looks like currently or, even worse, it may not sustain life as we know it at all, see (Fine-tuned universe, 2002).

There are numerous other theories hoping to encapsulate everything the standard model can account for and also include an explanation of all the remaining issues. Unfortunately, there is no single theory, to date, that has achieved this requirement. It was long thought that String Theory would indeed provide the ultimate theory of the universe. Some of the cleverest scientists in the world devoted decades of attention to its development. String theory was originally developed, between the 1970s and 1980s, as a theory of fundamental particles that did not necessarily involve gravity.

MODERN COSMOLOGY

Chapter 13
Cosmological Inflation

After the Big Bang event that started our universe, there was a brief period of rapid expansion. This violent expansion cannot be explained by the standard model. It is called cosmological inflation, and is theorized to have existed from 10^{-36} seconds, after the Big Bang, to around 10^{-33} seconds, when it ended. The universe continued to expand, but at a much slower rate compared to the incredible era in the evolution of the universe. Although brief in time, its presence had profound consequences for the remainder of the evolution of the universe to its current state. Moreover, it addresses the major issues outlined in the previous topic, above. Namely, the flatness and the fine-tuning problems. It also addresses the magnetic-monopole problem since it places a bound on their abundance. This makes it clear why efforts for their detection or observation have gone unsatisfied, see (Inflation (cosmology), 2022).

It is important to understand that the sudden expansion of the universe due to inflation is something distinct to its current accelerated expansion. Furthermore, there is no direct proof that inflation caused the acceleration process that scientists observe today. In part, this is because not enough is known about the details of the inflation process, but also that the two events occurred at vastly different times. On one hand, inflation initiated and completed within a minuscule amount of time in the early universe; the rapid expansion was a monumentally violent event. On the other hand, the acceleration of the expansion of our universe initiated around 7 to 8 billion years into its evolution, see (Carroll et al., 2017).

Cosmological inflation has a large following among cosmologists, with the vast majority believing it can lead to explanations of numerous issues in modern cosmology. This includes the large-scale structure of the universe and galaxy formation, as previously discussed. The details of the idea are actually based on quantum mechanics. This is curious because everything that we have discussed so far, like the standard model, has been rather classical in nature. Einstein's theory of General Relativity, for example, is a classical theory. It only deals with large scale structures in the universe. On a smaller scale, there is simply a continuum. That is, spacetime is smooth on all scales including the microscopic and, the even smaller, the Planck scale, see (Inflation (cosmology), 2022).

For example, the Planck length, denoted by **[Jacob 054]**, is defined in terms of three fundamental constants of Physics, see (Carroll et al., 2017). Namely: the gravitational constant, the speed of light and the (reduced) Planck constant. It has the definition and value given by

$$l_p = \sqrt{\frac{\hbar G}{c^3}}$$

where G is the gravitational constant, c is the speed of light and is the (reduced) Planck constant. The values of these fundamental constants are

$$G \approx 6.674 \times 10^{-11} m^3 . kg^{-1} . s^{-2},$$

$$c \approx 300000 km. s^{-1},$$

$$h \approx 6.582 \times 10^{-16} eV. s$$

where $\hbar = \mathrm{h}/(2\pi)$ is the reduced version of the Planck constant, denoted by h. The approximate numerical value of the Planck constant is therefore found to be

$$l_p \approx 1.616 \times 10^{-35}$$

where the units of this constant is in the SI units of meters. SI units (International System of Units) are standardized units of measure that have been agreed upon universally, see (Carroll et al., 2017).

This is an exceedingly small distance, and it would be inconceivable for most people to come to terms with given its infinitesimal scale. The Planck scale is not something of importance in everyday life nor even a fundamental constant, for example, like the speed of light. However, it is of great importance in physics and especially quantum mechanics, where it becomes most relevant. It plays an intrinsic role in the quantization of fundamental particles, such as photons, which are particles of light. It must also play a role when considering the quantization of spacetime. This is required if a full theory of quantum gravity is ever going to be achieved. If this is achieved then perhaps cosmological inflation, assuming it is correct, could be fully understood, see (Planck constant, 2022).

Cosmological inflation is clearly the forerunner as far as explanations of the evolutionary process that the early universe underwent. Aspects of this idea have been confirmed by observation, for example see (Tsujikawa, 2003). Many scientists believe this idea, or some version of it, is the correct answer, but there are a select minority of influential researchers that are not of the same opinion. Most notably Paul Steinhardt, Albert Einstein Professor in Science at Princeton University and director of the Princeton Center for Theoretical Science. Paul Steinhardt's opinion carries a lot of weight, given that he was one of the original contributors of cosmological inflation theory. He has effectively turned his back on the idea completely, in favor of something radically different.

Paul Steinhardt and his co-workers do not believe that the early universe underwent the process of inflation. They propose the revolutionary idea that there was no Big Bang event at all. But, rather, that our universe began as a bounce from a previously contracting universe. This of course presupposes another universe to have existed before ours: an idea that most scientists do not support or, at least, find hard to believe, as it is difficult to prove experimentally. He has also pointed out that predictions based on inflation are not consistent with the pattern of temperature variation observed in the cosmic microwave background (CMB). Another piece of evidence against inflation is that primordial gravitational waves, a relic of inflation, have never been detected, for example see (Corneliussen, 2017).

> *"The cosmology community has not taken a cold, honest look at the big bang inflationary theory or paid significant attention to critics who question whether inflation happened. Rather cosmologists appear to accept at face value the proponents' assertion that we must believe the inflationary theory because it offers the only simple explanation of the observed features of the universe."*,
>
> — Anna Ijjas, Paul J. Steinhardt, Abraham Loeb. (2017). "Cosmic inflation theory faces challenges," Scientific American.

Roger Penrose, like Paul Steinhardt and his co-workers, is also a staunch opponent of inflationary cosmology, and he also takes things a step further in his book (Penrose, 2016).

CHAPTER 14
IS THE UNIVERSE CLASSICAL OR QUANTUM

The standard model of our universe is based on a classical theory. Specifically, it is founded on Einstein's theory of General Relativity, which is a classical theory. Classical theories are built on the assumption that space, and time, or more precisely spacetime, is continuous or smooth at all scales. This means, if you take a powerful enough microscope and inspect an infinitesimally small part of spacetime, it will appear as smooth as a large part of space. This is in complete contrast to quantum theory where everything is discrete at a sufficiently small scale (the so-called Planck scale), discussed previously. However, as we have mentioned previously, the standard model falls short in answering pivotal questions about several issues, foremost the rapid expansion of the early universe. The commonly acceptable approach is to explain this phenomenon by the concept of cosmological inflation. However, inflation itself is thought to only be explained on a quantum level, although this is not yet fully understood, see (Powell, 2015).

Quantum effects are generally confined to small scales but can have profound consequences for our macroscopic universe. One of the most important examples of this involves cosmological inflation, which is assumed to be founded on quantum mechanical ideas. However, it should be stressed that the exact details behind the phenomena are still unknown. It may well keep eluding scientists until a full theory of quantum gravity is found, see (Quantum, 2022).

The mechanism of inflation, it seems, can only be explained at the quantum level, given the enormous energies involved in the early universe. All attempts, thus far, of producing a successful model have failed to be experimentally confirmed. This has certainly been the case with String Theory. This was recently evident at the Large Hadron Collider (LHC) experiments where there is an ongoing search for new particles predicted by the theory with so-called supersymmetry, see (Supersymmetry, 2022). No such particles have been found, apart from the Higgs boson previously mentioned.

On the other hand, Roger Penrose's Conformal Cyclic Cosmological (CCC) is totally based on a conformal extension of General Relativity, (Penrose, 2012). There is no appeal to the quantum realm at all, not even at the Big Bang event in each aeon. This suggests the universe, and all past and future aeons, are totally modeled by classical theory. There is no need to involve cosmological inflation as the rapid expansion of the universe is built into the model itself, see (Conformal Cyclic Cosmology, 2022). CCC is a cyclic universe theory in which a universe consists of an infinite succession of aeons where each aeon originates with its own Big Bang and has an exponentially expanding future, consistent with a positive cosmological constant.

It appears we are currently at a crossroad in cosmology. Only time will tell if we can achieve a total understanding of the universe using our classical or quantum way of thinking, see (Becker, 2022).

Chapter 15
Quantum Gravity

Two of the most successful theories ever conceived by the human mind, in an attempt to understand the nature of the physical universe, are Einstein's General Theory of Relativity and Quantum Mechanics, see (General Relativity, 2022) and (Quantum Mechanics, 2022). Relativity theory deals with the very large-scale, like galaxies and star systems. In addition to places where gravitational attraction is significant, like near a Black Hole. On the other hand, Quantum Mechanics is only relevant on the very small-scale and is one of the most accurate theories in the history of science. One is then tempted to propose a unification of these two theories. This would be useful since they both successfully explain the fundamental building blocks of the universe on opposite scales. Why not join them together and simply have one theory that covers all possible scales. Such a theory was named quantum gravity, see (Wood, 2019). It would supersede the classical picture of the universe and operate from the tiniest scale (the Planck scale) to the scale of the universe itself.

Well, when scientists attempted unifying General Relativity with Quantum Mechanics, they kept encountering mathematical singularities. Mathematical singularities occur when the mathematics of the theory 'misbehaves', typically by generating infinitely large values. These were similar to the singularities introduced previously for Black Holes and the Big Bang event. Encountering mathematical singularities when attempting to construct an emergent theory from two well-known ones is a bad sign. In fact, it indicates that either one or both theories are incorrect. This is indeed puzzling since both theories are so successful individually. It was quickly realized that this approach to constructing a quantum theory of gravity would simply not work. Most certainly, scientists have been at a loss to identify the exact problem with either General Relativity or Quantum Mechanics. General Relativity has a long list of successful predictions, including the existence of Black Holes and gravitational waves. While many scientists believe that quantum mechanics is the most accurate theory ever conceived.

> *"Somehow a theory of quantum gravity would need to reconcile these ideas about space and time. One way to do that would be to eliminate the problem at its source, spacetime itself, by making space and time emerge from something more fundamental."*,
>
> Becker (2022).

It was clear that a completely new approach was needed, and a number of promising candidates were proposed. This includes so-called loop quantum gravity, string theory, and twistor theory. Of course, this list is by no means exhaustive; there have been many attempts at a theory of quantum gravity. However, these three seem to be the leading contenders. Let us briefly explore each of these theories and consider how far they have progressed since their conception.

The first, loop quantum gravity, is credited to Professor Abhay Ashtekar and dates back to 1986, see (loop quantum gravity, 2022). His idea was to think about Einstein's Theory of General Relativity in the language of particle physics. The general construction consists of tiny little loops in the geometry of spacetime; you can think of this as a tapestry of sorts. A blanket is woven by a finite number of little woolen loops that are interconnected with each other. The blanket represents the geometry of spacetime, so you can image spacetime is made up of Planck scale loops. When used to understand the universe, loop quantum gravity or more specifically a branch called loop quantum cosmology, retains the Big Bang scenario of the standard model, (see, Illustration 16.). However, it also predicts a big crunch and a big bounce, see (Bojowald, 2008). The general idea is that the universe moves through a cycle. First, a big crunch, reducing the universe down to a singularity; producing a bounce and then expands again; only to be followed by another big crunch later. This cycle continues in this fashion.

Loop quantum cosmology, or more generally loop quantum gravity, appears to be one of the front runners in the quest to achieve a full theory of quantum gravity of the universe. However, even this proposal, despite its many successes has some fundamental issues to still iron out. For example, you can imagine that any successful theory of quantum gravity must be able to recover General Relativity, under certain limiting assumptions: that is, when it is assumed that only classical physics is relevant. It makes sense that under such an assumption the classical theory of gravity ought to be recovered. You may recall this is exactly what happens in the Newtonian case, as previously discussed. Einstein's theory of General Relativity was, in part, so successful because it was established that in the non-relativistic limit Newton's theory was recovered. The same type of result is needed from any proposal of a theory of quantum gravity, see (Loop quantum gravity, 2022).

> *"Loop quantum gravity describes spacetime as noncontinuous: instead of being smooth [as it is in Einstein's theory of General Relativity], it is broken into chunks or "atoms" of spacetime if you zoom in close enough. These building blocks give rise to what we experience as continuous spacetime through one-dimensional sheets that connect the blocks. These links create what physicists call a "spin foam".",*
> Becker (2022).

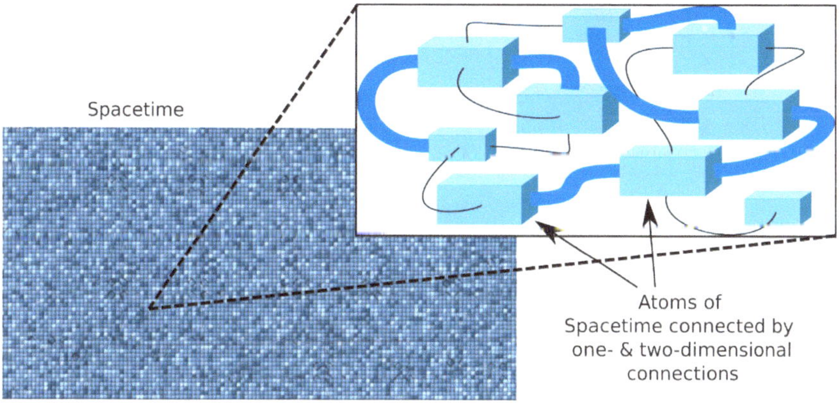

Illustration 16: Loop quantum gravity represents spacetime as being composed of 'Atoms', if one zooms in close enough, these Atoms are connected with each other through one-dimensional strings and two-dimensional sheets. They form what is referred to as spin foam or quantum foam.

The second, string theory, is credited to a large group of scientists that have pieced together the theory since the late 1960s, see (String theory, 2022).

The theory started out from a particle physics perspective and then included gravity as an additional component later. The general idea is that all so-called fundamental particles are not thought of as particles, but rather as objects called strings. A simple string, that is either open ended or closed ended, exists in one-dimension and can represent all fundamental particles by the frequency of its vibration. This is indeed a fascinating idea and has led to a revolution in pure mathematics, but unfortunately not in physics.

String theory is founded on a remarkable piece of theory called supersymmetry. Many scientists believe that supersymmetry must be true in the fundamental description of particle physics. This theory connects bosons and fermions, introduced previously, in a fundamental way. But here is the major problem: string theory, together with supersymmetry, predicts that extra particles should exist. This was seen as a way of putting string theory to the experimental test. The energies at which these extra particles are predicted to exist is low enough for a particle collider to recreate. Particle colliders, like the one at the European Organization for Nuclear Research (CERN), can recreate particles through high energy collisions. If fact, CERN is the site of the world's largest and highest-energy particle collider called the Large Hadron Collider (LHC), see (CERN, 2022). However, in the late 2010s, when scientists ran experiments to find these extra particles, no new fundamental particles were discovered. This was seen as a major fail for string theory, if not a significant setback. Since that time, the research output for string theory, as made available on the pre-print archives, has exponentially decreased. We will have to wait until there is further development at CERN on this topic.

The third in the list of attempts at developing a theory of quantum gravity is Roger Penrose's twistor theory, see (Twistor theory, 2022). As with both loop quantum gravity and string theory, twistor theory has branched out into other areas of mathematics and physics. It has proven useful in answering many theoretical questions. Unfortunately, with respect to being considered as a full theory of quantum gravity, it has hit a brick wall. Referred to as the gravitational googly problem, this intellectual brick wall has halted any progress for over 40 years, see (Penrose, 2015).

So, what do we have, after serious efforts and contributions from some of the most gifted minds in the world for over half a century? Well, at best, it is a fragmented understanding of what a theory like quantum gravity might look like. This is indeed very disappointing, not only when considering the many years spent on this problem, but also the many millions of dollars spent attempting to find confirmation experimentally, for example, in the form of particle accelerators like the one at CERN.

Chapter 16
The Cosmic Microwave Background

One of the greatest achievements of modern physics is not only the accidental discovery of the cosmic microwave background, but also its use that led to the Big Bang model of the universe. The cosmological microwave background (CMB) is an historic electromagnetic radiation which has been left over from the early universe, see (Durrer, 2015). It is one of the key pieces of evidence scientists have to support the Big Bang hypothesis. Historically, its discovery in the mid-1960s eventually gave rise to the idea of an evolving universe over the long-held belief it was in a steady-state. In fact, Einstein himself adapted the steady-state model and, as a result, introduced the cosmological constant into his formulas of General Relativity.

It is of fundamental importance for experimental data because it is the oldest observable electromagnetic radiation in the universe. It's not the type of thing you see with your naked eye, but rather with the use of radio telescopes. These sensitive instruments record a background microwave signal that permeates throughout space. A remarkable feature of the cosmic microwave background is that it is almost isotropic. It cannot be attributed to galaxies, planets or even stars. It comes from the past, when the universe was in its early stages of evolution. Therefore, it gives scientists great opportunity and has led to significant insights into the evolution of our universe, see (Cosmic microwave background, 2022).

Robert Wilson and Arno Penzias, two radio astronomers, won the 1979 Nobel Prize for discovering the cosmic microwave background. They spent many months trying, in vain, to get rid of a background noise during their tests with a large horn reflector antenna. Their original experiment was focused on measuring faint radio waves bouncing off satellites orbit around Earth. They first attributed this noise to a list of possible local sources, like pigeons, defects in the antenna surface, and even settlement from dirt and debris, local sources of external noise and vibration, like traffic etc. But no matter what they tried, there was a persistent noise in the background of all their measurements. It seems to come from all directions and had the two radio astronomers baffled. After extensive investigations, it was finally revealed that what this noise was, emanated from the universe as so-called black-body radiation. They even estimated its temperature at around 3 Kelvin (approximately -270 degrees Celsius), (see, Illustration 17.). This discovery quickly attracted vast attention to the Big Bang model but also ushered in the end of the steady-state model era, see (Discovery of cosmic microwave background radiation, 2022).

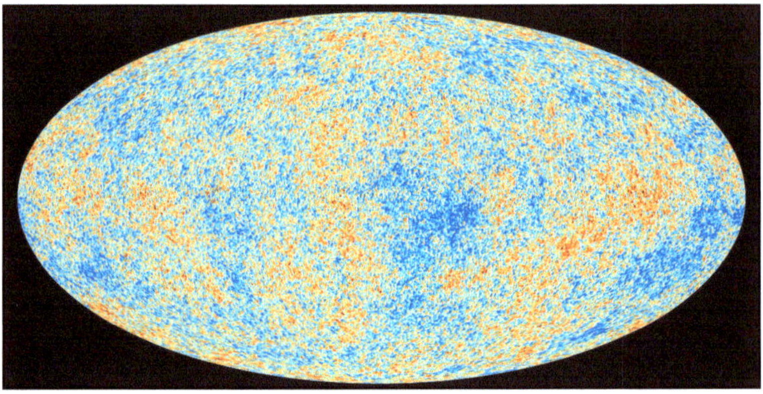

Illustration 17: The cosmic microwave background, visible in all directions, shows the hot and cold patterns observed. Slight temperature differences on the background sky left billions of years ago when our universe first became transparent to light. Image Credit: European Space Agency, Planck Collaboration. https://apod.nasa.gov/apod/image/1303/cmbr_planck_3600.jpg

FUNDAMENTALS OF DARK MATTER

CHAPTER 17
HAWKING RADIATION AND DETECTING DARK MATTER

As discussed previously, Stephen Hawking discovered a remarkable fact about Black Holes – they radiate. Up until that time, mid-1970s, it was understood that Black Holes were simply 'black holes', nothing was thought to be able to escape. The gravitational pull of a Black Hole is so strong that when you reach the point of no return, something called the Schwarzschild radius (recall from Chapter 4: Basics of Thermodynamics: Entropy, paragraph 6 this is the point of no return called the 'event horizon'), you can never escape. However, Hawking proved this is not necessarily the case. He combined, or unified, three theories of physics at a place called the Event Horizon of a Black Hole, see (Black hole thermodynamics, 2022).

As we have previously discussed, stellar-mass Black Holes, and certainly larger Black Holes, will not totally evaporate until many more years into the evolution of the universe because Hawking radiation is so miniscule. So, there is little hope of detecting such Black Holes based on Hawking radiation. This argument no longer holds for Primordial Black Holes since, as we have shown, their mass can be as tiny as the Planck mass. It follows that these types of Black Holes can evaporate well within the age of our universe. This is true if their mass is below 10^{11} kilograms, as shown previously, see (Carroll et al., 2017).

It was Stephen Hawking, together with Bernard Carr, who first proposed the idea that Primordial Black Holes are behind the elusive missing mass of our universe called dark matter. This idea did not seem to gain much interest in the scientific community because a Primordial Black Hole has never been detected. However, this was also true of Black Holes themselves, and they were eventually detected. This success was considered one of the greatest achievements of theoretical physics. There is no reason to rule out the existence of Primordial Black Holes simply because they have not been detected. Primordial black holes could finally be detected using the new James Webb Space Telescope. This powerful telescope was launched December 25th, 2021. It was completely unfolded on January 8th, 2022. It began recording images in July 2022, see (NASA, 2022) (Illustration 18) see (NASA, 2022).

> *"If the first stars and galaxies already formed in the so-called 'dark ages,' Webb should be able to see evidence of them,"*
>
> – Professor Günther Hasinger,
> ESA's director of science, see (Mamoun, 2021).

Let us briefly consider the way the James Webb Space Telescope will navigate the solar system, as this is a fascinating process, see (Harwood, 2022). From a celestial mechanics perspective, scientists had to consult a piece of mathematics that has been developed since the late 1700's called the three-body problem. In this particular case, the three bodies are the Earth, the Sun, and the satellite itself. This is a special type of three body problem because the mass of the Earth and Sun are far greater than the mass of the satellite. In this ternary system, there are five special regions of space called Lagrange points (named after the Italian mathematician Joseph-Louis Lagrange), and denoted by the labels L1, L2, L3, L4, and L5, (see, Illustration 19). These points are considered special because gravity tends to have relativity little influence in these regions. This means that objects like satellites near a Lagrange point require less energy to stay there, compared to any other region of space.

Illustration 18: James Webb Space Telescope's stunning first full-color images and spectroscopic data, released in mid. July 2022. Credit: NASA, ESA, CSA, and STScI. https://www.nasa.gov/webbfirstimages

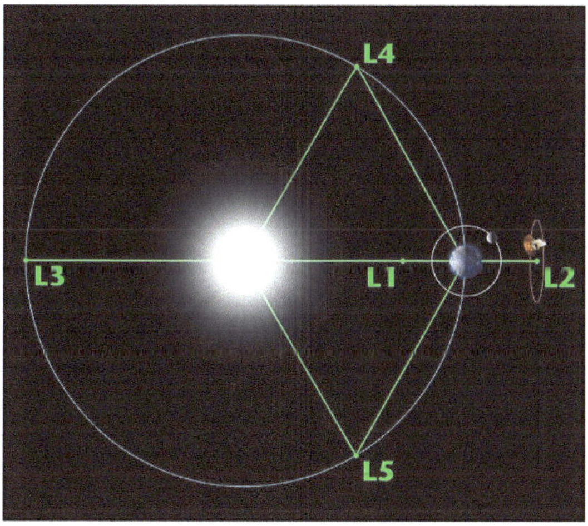

Illustration 19: The Lagrange points associated with the Earth, Sun, and satellite system. WMAP orbits around L2, which is about 1.5 million km from the Earth. Lagrange Points are enhanced regions of gravitational attraction and repulsion. These points are used by spacecraft to reduce fuel consumption that is needed to remain in position. Credit: NASA / WMAP Science Team. https://map.gsfc.nasa.gov/media/990528/index.html

Chapter 18
The Mystery of Dark Matter and Dark Energy

The amount of mass and energy in our universe is finite and can neither be created nor destroyed. It was created at the Big Bang and has, since that time, continued to expand, albeit remaining finite in total. Therefore, we should be able to account for all the mass and energy in the observable universe. Strangely, this is not the case, and the amount of mass and energy we can observe or account for is tiny. This has forced scientists to call the 'missing' matter and energy, dark matter and dark energy, respectively, see (Clegg, 2019).

Dark matter is attributed as making up 23% of the matter in the universe. This is a monumental amount that cannot be accounted for through observation. This means the amount of matter we can see in the observable universe is only 5%. The 23% of missing matter either cannot be observed directly, it is invisible, or there is something fundamentally wrong with our understanding of the universe. Our current understanding of the universe requires this missing amount of matter, dark matter, in order to satisfy the requirements of our models given the observations we make. The addition of dark matter to these models aligns what is observed with the predictions of the modeling, see (Dark matter, 2022).

Dark matter is called 'dark' because it cannot not be observed like ordinary matter. Ordinary matter interacts with electromagnetic radiation. This means, for example, that everyday typical objects like teacups, doors, trees and so on, all absorb and emit electromagnetic radiation. Your eye, the detector, absorbs this radiation in the form of visible light and then your brain processes the information. In the case of dark matter, it would simply be invisible to you because no electromagnetic radiation is being emitted or reflected, see (Clegg, 2019).

According to our current understanding of the universe, based on the standard model (referred to as LCDM (Lambda-Cold Dark Matter), not to be confused with the standard model of particle physics), the non-inclusion of dark matter would lead to catastrophic events. Without the additional gravitational attraction supplied by the presence of dark matter, it would be difficult for galaxies to form. If they did manage to form their constituent matter, planets, stars and so on, could easily scatter apart due to dynamical instabilities from insufficient gravitational attraction.

There is convincing experimental data strongly suggesting that there is indeed missing matter in the universe, that is dark matter, independent of our standard theory and models. Recently, NASA released a report on the direct detection of dark matter, (Hupp et al., 2006). By using NASA's Chandra X-ray Observatory and other telescopes, they observed a phenomenal collision between two large clusters of galaxies. These are known as the galaxy cluster 1E0657-56 or, more colloquially, the bullet cluster, named after its spectacular bullet-shaped cloud of hundred-million-degree gas.

> *"This is the most energetic cosmic event, besides the Big Bang, which we know about,"*
>
> — Maxim Markevitch of the Harvard-Smithsonian Center for Astrophysics in Cambridge, Mass.

Using telescopes around the world and even the Hubble Space Telescope, they were able to calculate the location of the mass in the clusters. This calculation relied on the effect of gravitational lensing, as previously discussed, (see, Illustration 20). This is an analytical tool that was known to Einstein himself and was born out of his theory of General Relativity. It works by measuring how gravity from the clusters of colliding galaxies distorts light from background galaxies. The hot gas of the bullet cluster experienced a force like air resistance during the collision and, as a result, its velocity started slowing down. On the other hand, the dark matter that was present did not slow by the impact, see (Gravitational lens, 2022).

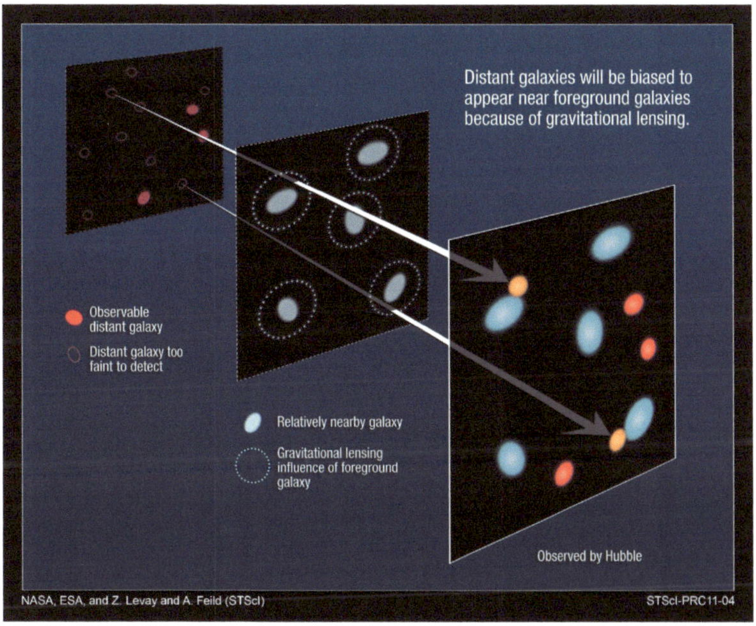

Illustration 20: Illustration of how gravitational lensing by foreground galaxies will influence the appearance of far more distant background galaxies.. Illustration credit: NASA, ESA, and A. Feild (STScI). https://www.nasa.gov/images/content/509785main_hs-2011-04-a-print.jpg

This was the 'smoking gun' that scientists were looking for because it implied a separation of ordinary matter and dark matter. The dark matter only interacts with the gas of the clusters through gravity and no other forces. On one hand, the gas was being dragged on by a resistive force and, on the other hand, the dark matter simply ignored this dragging force. There was a clear separation of the dark matter and ordinary matter as measured in the data from the telescopes. This required only the theoretical framework of General Relativity and the concept of gravitational lensing, which is well established, highly reliable and accurate. If there was no dark matter and the gas was the only type of mass involved, then such a separation could not be possible.

This result also spells doom for alternative or modified theories of gravity that are based on the premise that Einstein's theory of General Relativity is somehow incorrect. A common statement is proclaimed: that General Relativity must be modified for large scale interactions associated, for example, with galaxies. The results above clearly demonstrate that the center of mass for the Bullet cluster is displaced considerably from the dark matter center of mass.

There is a clear separation that cannot be explained by any other, alternative or modified, theory. This powerful result verifies that indeed, dark matter is missing mass – it is not a consequence or deficiency of any theoretical model.

> *"This is the type of result that future theories will have to take into account, …, As we move forward to understand the true nature of dark matter, this new result will be impossible to ignore."*
>
> Professor Sean Carroll,
> American theoretical physicist.

There is the related concept of dark energy. According to the standard model, dark energy makes up 72% of the energy in the observable universe. Unlike dark matter, it is somewhat more hypothetical in nature. However, just like dark matter, it only seems to interact through gravitational attraction alone. Although it makes up a large portion of energy, scientists believe it has low density and is smeared uniformly in vacuum space. Of course, this suggested that observing it through experiment or measurement like dark matter would be unlikely. In fact, all attempts to experimentally detect dark energy have failed, see (Sabulsky et al., 2019).

The name dark energy is just a modern renaming of the so-called cosmological constant, see, for example (Lahav, 2020). The cosmological constant was introduced by Einstein himself to model a steady-state universe, see, for example (O'Raifeartaigh, 2017). This turned out to be a monumental mistake, with Einstein calling it his greatest blunder. He made this statement because it was later found that not only was the steady-state model superseded, but also that his attempt to model it failed. That is, without the cosmological constant, he could well have predicted the Big Bang model. Surprisingly, the cosmological constant turned out to be of great relevance anyway but just not for its original intention. Ironically, even Einstein's blunders turn out to be great ideas after all, see (Dark energy, 2022).

Let us think about what Einstein attempted to do by introducing this extra term into his equations of gravity. He effectively attempted to use dark energy to balance out gravitational attraction. This is a great example of what we mean by 'fine-tuning'. He tried to fine tune his model to fit with what he, and many others, thought was the correct theory at the time. Einstein believed that the role of dark energy was to act as a type of gravitational 'negative mass' that would be distributed throughout vacuum spacetime, that is, empty space. This idea has been verified within General Relativity, but just not in the same way as originally intended.

In the present age of the universe, it is estimated that, there is 72% dark energy and 28% dark matter, of which only ~5% is made up of atoms, that is, ordinary matter. In particle physics, ordinary matter is made of so-called baryons, introduced previously. Most of the matter of the universe is baryonic matter. This includes most atoms with the property of mass. On the other hand, non-baryonic matter is any other matter not composed of baryons. For example, neutrinos, electrons and, of course, dark matter and black holes (see, Illustration 21).

Dark matter can be sorted in two possible types: cold dark matter (CDM) and/or hot dark matter (HDM). Of the two possibilities, cold dark matter is the favorite candidate. This is simply because hot dark matter is associated with relativistic speeds (a large fraction of the speed of light). Such particles are less likely to lead to clumping, in the early universe, due to gravity. Hot dark matter is typically thought to be composed of massive leptons (they are non-baryonic, of course). In the case of the standard model only cold dark matter is considered; it is actually in the name Lambda-**Cold-Dark-Matter** (L**CDM**). Cold dark matter is thought to be composed of various candidate exotic particles that have never been previously detected. Examples include so-called WIMPs (weakly interactive massive particles) and axions that are bosons. These have never been detected. Over the years, there have been many propositions to explain the existence of dark matter and dark energy. All of them appear to have an issue of some kind. There is no clear, definitive, agreed-upon explanation of the missing mass and energy that is apparent in our universe, see (Baryonic dark matter, 2022).

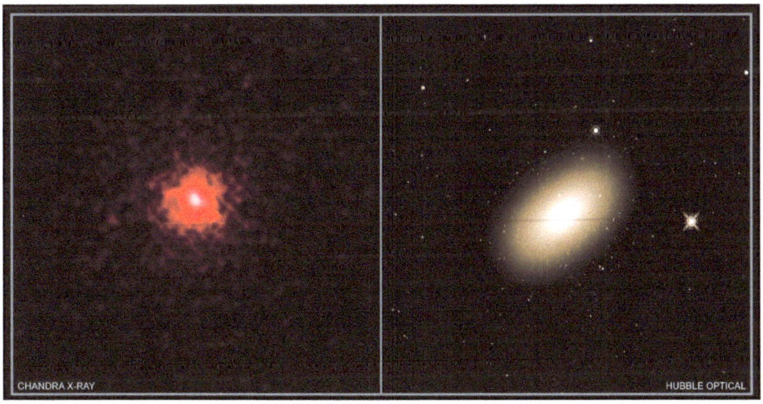

Illustration 21: The galaxy Markarian 1216 has more dark matter packed into its core than expected. Image credit: X-ray: NASA/CXC/Univ. of CA Irvine/D. Buote; Optical: NASA/STScI. https://www.nasa.gov/mission_pages/chandra/images/heart-of-lonesome-galaxy-is-brimming-with-dark-matter.html

CHAPTER 19
PRIMORDIAL BLACK HOLES AS DARK MATTER

A possible and plausible candidate for dark matter is primordial Black Holes. We previously introduced these types of the Black Holes (see Illustration 22). The important property they have is that there is no restriction on their total mass. In reality, we expect their mass to range from 10^{-8} kilograms to 10^5 solar masses. The only requirement for an object to be classified as a Black Hole is that it must be entirely encapsulated within the event horizon. Notice that there is no rule or requirement about how it was actually formed in the first place. The processes that lead to the formation of Black Holes are irrelevant. Only the end results are of any relevance when deciding if a celestial object is a Black Hole or not, see (Primordial black hole, 2022).

The main difference between a Black Hole and a Primordial Black Hole is the conditions under which they were created. A Black Hole must be formed from a collapsing star of sufficient mass. On the other hand, a Primordial Black Hole is conjectured to have formed in the early universe. Since there were no stars around at that epoch of the universe, the formation must have been achieved though gravitational clumping or coalescence.

This radically different idea for the constituents of dark matter was first introduced by Stephen Hawking, as discussed previously. Primordial Black Holes as dark matter candidates were initially ruled out because there has been no evidence such Black Holes exist. However, there are renewed efforts to determine if they actually exist, see (Mamoun, 2021). In the future, there is a possibility of their detection by the James Webb Space Telescope.

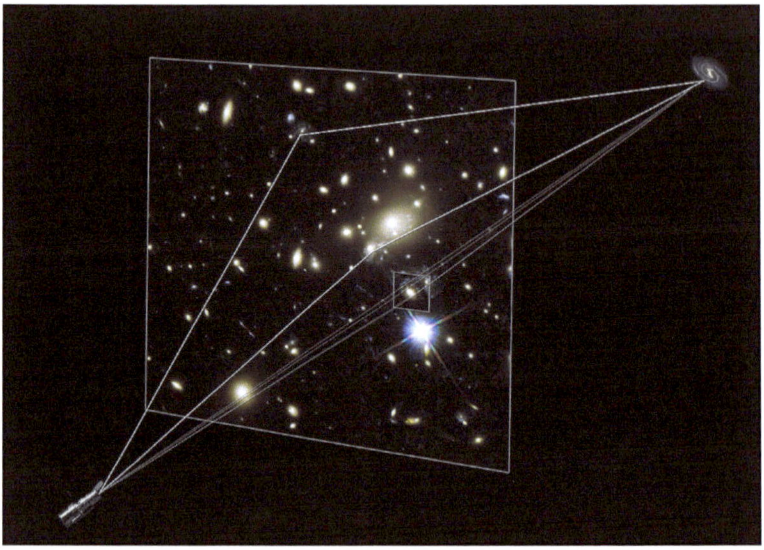

Illustration 22: Shining a light on dark matter with NASA's Hubble Space Telescope. Image Credit: NASA/ESA. https://www.nasa.gov/image-feature/shining-a-light-on-dark-matter

According to recent findings, (Fore, 2018), there is an intimate relationship between primordial Black Holes and Higgs bosons. Recall, we introduced Higgs bosons and their role in the early universe previously. It turns out that this so-called God particle may be unstable, and this leads to quantum fluctuations that produced highly dense conditions that support the creation of mini primordial Black Holes. So, let us breakdown what this all means. When we say a particle is unstable, this means that it is not yet in its final state. To reach a final state it must go through the process of particle decay, (Particle decay, 2022). This process leads to quantum fluctuations in the Higgs field: think of this like waves in the ocean on a stormy night, some areas of the field will be highly dense compared to others. So, you can think of these high-density areas as the peaks of each wave in the ocean. This high density was sufficient to cause it to collapse in on itself, thereby producing these so-called primordial Black Holes. The theory, (Fore, 2018), suggests these primordial Black Holes started with small masses in the order of 10^{12} kilograms. They subsequently clumped together as the universe evolved. All this leads to an extraordinary theory that dark matter is related to these primordial Black Holes.

> *"The theory is appealing on many levels: Dark matter does not emit detectable light; neither do black holes. And primordial black holes, if they exist, could be physical evidence that the Higgs field is not at its lowest possible energy, and is instead in a quasi-stable state. The new theory can also explain dark matter without requiring any new physics beyond the Standard Model"*,
>
> Fore, Live Science, see (Fore, 2018)

Recently, Clesse and Garcıa-Bellido (2018) listed seven hints that primordial Black Holes are indeed responsible for Dark Matter.

> *"The failure of a long series of direct detection experiments ... to detect a Weakly Interacting Massive Particle (WIMP) or any other particle candidate now leads some cosmologists to reconsider alternatives to particle Dark Matter models. One of them has received recently a lot of attention, after the detection of several black hole mergers by Advanced LIGO/VIRGO since September 2015 ... the possibility that Dark Matter could be constituted in part or entirely by Primordial Black Holes (PBH)."*,
>
> Clesse and Garcıa-Bellido (2018).

These authors give compelling evidence that primordial Black Holes are not only important in explaining the mystery behind Dark Matter but also that they contribute to the understanding of other processes.

> *"we claim that [Primordial Black Holes] PBH as cold dark matter would provide a natural explanation to the longstanding problems of cosmology and astrophysics, namely the existence of supermassive black holes at high red-shifts, the missing galactic satellites, the too-big-to-fail, the core-cusp problems, and the missing baryons in the Universe."*,
>
> Clesse and Garcıa-Bellido (2018).

There is now compelling evidence that primordial Black Holes are connected with dark matter in someway not fully understood.

> *"[T]he most exciting possibility is that primordial black holes are numerous enough to account for the dark matter. According to García-Bellido's model, the Milky Way Galaxy should be*

swimming in a sea of roughly a trillion black holes. "When you tell people that space may be full of black holes, they get nervous," García-Bellido says. But he adds that the separations are so large that the probability of a black hole coming near our Solar System is very small.",

Michael Schirber (Schirber, 2018)

A SIMPLE MODEL OF THE UNIVERSE

CHAPTER 20

A NEWTONIAN APPROXIMATE OF THE UNIVERSE

It should be clear that any calculation associated with the modeling of our universe is extremely complicated. This is even the case when the theory behind the calculations fails to adequately explain key features of the phenomena or experimental observations. Scientists worldwide rely on the foundations of the so-called standard model for, at least, a partial explanation of our universe. The standard model is founded on Einstein's theory of General Relativity. This theory is arguably the greatest theory of all time. It has stood the test of time, again and again. Recently, as another one of its powerful predictions, Black Holes were experimentally confirmed. The Nobel Prize in 2020 was in fact awarded to Sir Roger Penrose for rigorously proving the existence of Black Holes using the mathematical concept of trapped surfaces. (Later, his work with Stephen Hawking led to a generalization of these results.) Reinhard Genzel and Niklas Elmehed were also awarded the 2020 Nobel Prize for their part in establishing the experimental evidence.

Despite the fact of the many successes of the standard model, it also has many failures or insufficiencies, or so it would seem. For example, it cannot account for the inflationary epoch shortly after the Big Bang. The inflationary period is an essential element in the evolution of the early universe. It describes an exponential expansion that was far more rapid than can be explained by the standard model. The concept of inflation itself is founded on quantum mechanical considerations and is therefore beyond the scope of the standard model.

To model the entire evolution of the universe, even when confining ourselves to the standard model, is almost unthinkable. This is purely due to the complicated nature of the Einstein field equations that describe such an evolution. Mathematically, one would have to formulate what is known as an initial value problem and solve it exactly. This is a formidable task, even in very specialized cases involving simplifying assumptions that are aimed at reducing the complexity of the problem. For example, the field equations of General Relativity are highly non-linear such that even the most powerful supercomputers would not be able to solve them for this problem. If the next step is moving to a more complete theory, something that String Theory was hoped to fulfill, this task becomes insurmountable. Simply because the field equations themselves are difficult to write down, let alone to solve.

Sometimes, it is important to use approximate or phenomenological methods to indicate the general behavior of the physical system being modeled. These methods have proven invaluable in the absence of a complete theory or when faced with intractable mathematical equations. On the other hand, it is essential to realize the limitations of such approaches. A good degree of caution must be maintained when interpreting outcomes. Care must be shown in order not to interpret such results at face value. Regarding the modeling of the evolution of the universe, we face an immense mathematical obstacle, one that can be side-stepped using an appropriate Newtonian approach.

In the analysis that follows, we assume, for simplicity's sake, that an approximate approach to modeling the evolution of the universe is sufficient. In fact, we will consider a Newtonian framework for all epochs of the evolution of the universe. Such a simplistic picture cannot not possibly account for all details in the broader understanding of the universe. However, this approximate Newtonian approach will enable the construction of a 'hand waving' argument. Moreover, it will be easy to convey the basic ideas without resorting to complicated theories and formulae.

Suppose the total mass of the universe is denoted by M; then the kinetic energy denoted by E, of the universe would be

$$E = \frac{1}{2} M V^2$$

where M is the sum of the mass of dark matter with the mass of ordinary matter, and V is the initial velocity of expansion.

The initial velocity here is not actually the initial velocity associated with the Big Bang. Rather, it is the speed of separation given by Hubble's law

$$V = H_o D$$

where H_o is the Hubble constant and D is the 'proper distance' from one 'clump' of matter which can change over time (and can eventually form a galaxy). The 'proper distance' is roughly the distance from a far-off object at a specific moment of cosmological time, which can change over time due to the expansion of the universe.

The motion of astronomical objects, due solely to expansion, is known as the Hubble flow. This is an important law as it applies both to homogeneity (recall, from above, this meant all things look the same over time) and isotropy (again, from above, meaning all things look the same no matter your location). So, Hubble's law holds true on a fundamental level; it is a universal law.

We will use this, as we call it, Newtonian approximation approach in the following chapters.

CHAPTER 21
NEWTONIAN DUST MODEL OF THE UNIVERSE

The idea here is that the universe is modeled as being filled with pressure-less 'dust' of uniform density. When we say modeled, what is really meant is that this is a simple representation of all the complexities of the universe. All the irregularities made up from galaxies, Black Holes, Neutron stars supermassive Black Holes, stars and planets, and every other conceivable object, are approximated with a dust-filled model, where each constituent of the dust is identical, non-interacting and therefore pressure-less. By pressure-less we mean that each constituent of the dust doesn't interact with anything else. The assumption that it is uniform simply reflects the idea that there are no perceivable irregularities in the dust, that is, there is no clumping or holes. It turns out that for large-scale models, this is actually a reasonable approximation for the present-day universe. Of course, this becomes much less reasonable as we rewind the evolution of the universe to the Big Bang event. Nonetheless, to keep the model as simple as possible, it will be assumed that this dust model is valid throughout the evolution of the universe. It is also important to realize that there is only one type of constituent of dust. In reality, this could not possibly account for all the different elements that make up the universe, such as photons, matter particles, etc. However, it will be sufficient for our needs and will serve as a simple model to demonstrate broad ideas about the universe.

To construct this model of the universe we will follow (Carroll et al., 2017). This model is shown in Illustration 23. The uniformity of the model is represented by the density, $p = p(t)$ which is a function of cosmological time t.

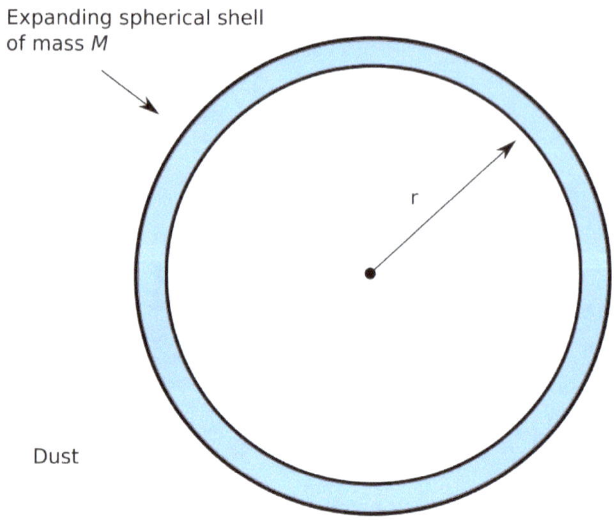

Illustration 23: Model of Newtonian approximation of the universe.

This model represents a homogeneous and isotropic universe. This is the reason we say it is a reasonable model for the present universe but not the early epochs of the universe.

CHAPTER 22

THE BIG CRUNCH

(NOT TO BE CONFUSED WITH THE "BIG CRUNCH" PROPOSED IN THIS BOOK)

There are three general possibilities to the ultimate fate of the universe. It may continue to expand at the current rate indefinitely and fizzle out. On the other hand, it may continue expanding until it reaches some stable point and then remains static. Finally, it may continue to expand to a certain point and then start to contract. This contraction would ultimately end with a Big Crunch implosion. The dynamics of this Big Crunch would basically be the reverse of the Big Bang. A final possibility, that has become popular in recent times, is the 'bounce' or 'pre-Big Bang' scenario. In this idea the universe undergoes a collapse or crunch, of sorts, thereby giving birth to a new universe through the Big Bang process, see (Betz, 2021).

Why, one might ask, can we be so confident that the Big Crunch would follow? (See, Illustration 24). Well, if all the Primordial Black Holes were 100% dark matter, then the universe would collapse on itself and there would be no further expansion from the Big Bang at all. Black Holes are truly the masters of the universe and have shepherded the galaxies into filaments.

In the coming chapters we will explore this idea further and in more detail.

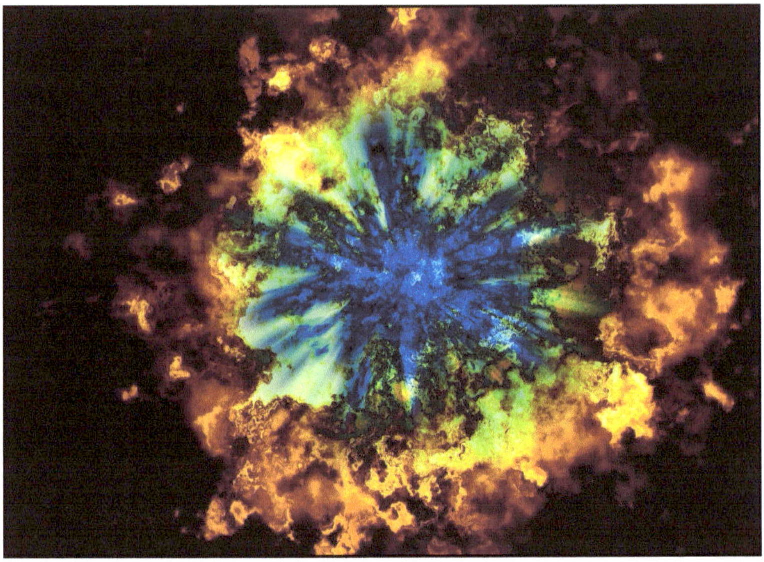

Illustration 24: Artist's illustration of the Big Crunch. Image Credit: https://cdn.pixabay.com/photo/2014/08/20/05/23/armageddon-422308_1280.jpg

PRIMORDIAL BLACK HOLES AND THE BIG CRUNCH

CHAPTER 23
PRIMORDIAL BLACK HOLES AS DARK MATTER IN THE UNIVERSE

Two massive black holes that merged some 1.3 billion years ago released so much energy that this event was detected as gravitational waves in September 2015 by LIGO (The Laser Interferometer Gravitational Observatory), see (Commissariat, 2016). This was the first time that gravitational waves were observed and was as predicted by Einstein's theory of General Relativity. Since then, many more mergers of such massive black holes have been observed, see (Illustration 25). This has spurred many cosmologists to renew their quest to prove that Primordial Black Holes exist and could account for most if not all dark matter.

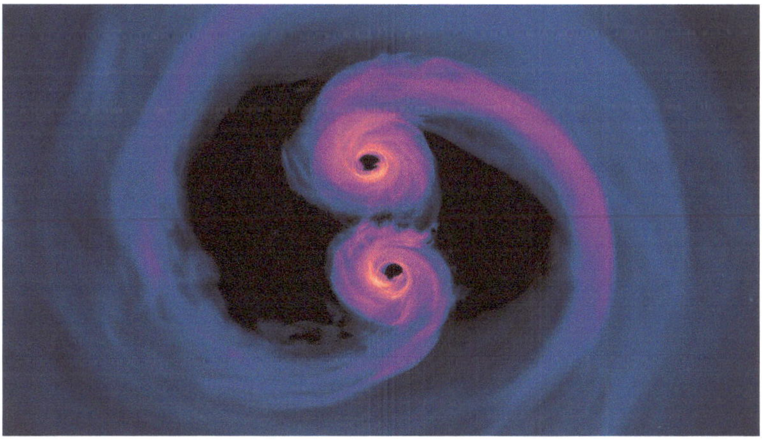

Illustration 25: Computer simulation of merging supermassive black holes producing gravitational waves. Credit: NASA's Goddard Space Flight Center. https://svs.gsfc.nasa.gov/13043

Should Primordial Black Holes be responsible for most of dark matter then this could result in a huge contraction in the velocity of expansion of the universe in the first second after the Big Bang. This contraction would be so big that it would rival the Big Bang itself and the kinetic energy of the Big Crunch would be stored as potential energy. In time this potential energy would be released into the universe and would be perceived as the universe speeding up. The energy of this contraction is shown to be 72% of the total energy in the universe, which is the exact number that is thought to be dark energy and is the core essence of this book.

An approximate curve for the expansion of the universe is developed. It helps explain why the universe would speed up after about 4 to 6 billion years, when most of the supermassive and other black holes would have formed. Eventually the expansion of the universe would approach the initial velocity of expansion. Should this model of the universe be verified we would no longer need to refer to dark matter and dark energy. Dark matter would simply be Primordial Black Holes and the dark energy would be the potential energy stored in all Black Holes and ordinary matter from the Big Crunch. Although not proven, cosmologists should pursue this, or similar concepts, as it would have far reaching consequences in explaining the universe as we know it today.

Primordial Black Holes are objects formed just fractions of a second after the Big Bang and are the leading contender in explaining the nature of dark matter. The advantage of such an assumption is that we do not have to include new forms of matter to explain dark matter in the universe such as WIMPS (weakly interacting massive particles). Thus far there is no evidence found to say that WIMPS even exist.

Recently several authors have proposed that Primordial Black Holes born in the first one second after the big bang could explain dark matter. The landmark article by Juan Garcia-Bellido and Sebastian Cleese, see (Garcia-Bellido and Cleese, 2017), was published in "Scientific American (July 2017)" and goes into a detailed explanation of how these primordial black holes could be formed due to density fluctuations in the primordial soup.

Camille Carlisle, see (Carlisle, 2019), in the April 3, 2019, issue of "Sky and Telescope", has also proposed that dark matter could be black holes. In that article, it is proposed that primordial black holes could be born in the earliest age of the universe when the cosmos was nothing but a hot plasmatic soup. This radiation was not uniform, and its density fluctuated from patch to patch. If a patch was excessively dense then it would naturally collapse and form a black hole.

In September 2015 it was determined that 1.3 billion years ago two black holes spiraled around each other and when they merged the energy released was so large (about four mass equivalents of our sun) that it shook the fabric of spacetime and was detected by sensors of the Laser Interferometer Gravitational-Wave Observatory (LIGO), see (Illustration 26).

Since then, a significant number of these large black hole collisions have been observed and adds credence to the fact that these black holes predate the formation of stars themselves. The sizes of these black holes were slightly larger, and their rotations rates were slower, than expected from stellar models. The chart in their article shows the masses of the five confirmed black hole mergers.

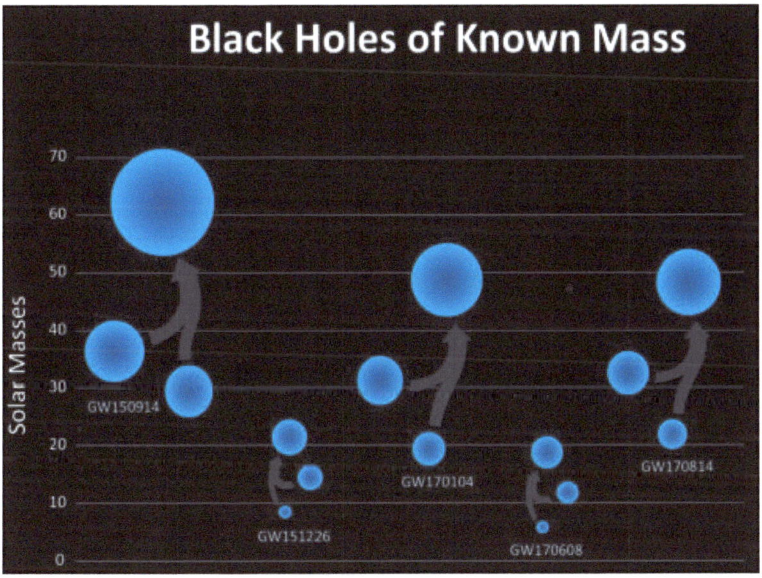

Illustration 26: This graph shows the Five Black Hole Mergers discovered using gravitational waves, according to their mass in Suns. LSC/ LIGO / Caltech / Sonoma State (Aurore Simonnet)

A recent publication by Christopher Berry, a member of the LIGO team, see (Berry, 2019), in "Scientific American, Dec 2019", says that since the original detection, sensors have detected 43 more events, making them seem almost commonplace and lends support to the frequent existence of these massive black holes.

In their 2017 article, in "Scientific American", Garcia and Cleese state that "Primordial Black Holes could constitute some, if not all of dark matter, the invisible 85% of the matter in the universe that acts as gravitational glue to hold galaxies and galaxy clusters together". More recently, see (Garcia-Bellido and Cleese, 2018), in "Physics of the Dark Universe", volume 22, Dec 2018, they cite that there are seven observations that point towards the existence of Primordial Black Holes, constituting all, or an important fraction, of the dark matter in the Universe.

Riccardo Murgia, see (Murgia, 2019), lead author of a study published in Sep 17, 2019, has also looked into the theory that Primordial Black Holes could be the principal candidates in explaining the nature of dark matter. His study was also spurred on by the gravitational waves seen by LIGO.

Earlier, in May 2016, Alexander Kashlinsky an astrophysicist at NASA, see (Reddy, 2016), suggested a possible link between Primordial Black Holes and Dark matter. He stated,

> *"These studies are providing increasingly sensitive results, slowly shrinking the box of parameters where dark matter particles can hide, … The failure to find them has led to renewed interest in studying how well primordial black holes — black holes formed in the universe's first fraction of a second — could work as dark matter.",*
>
> – Alexander Kashlinsky, (Reddy, 2016).

An intriguing alternate view is that dark matter is made up of black holes that form during the first second of our universe's existence. This interpretation aligns with our knowledge of cosmic infrared background (CIB) and cosmic X-ray background (CXB) glows. They may explain the unexpectedly high masses of Black Holes detected by LIGO in 2015. If this is correct, then all galaxies including our own are embedded with a vast sphere of Black Holes each with a mass about 30 times our Sun's mass.

In 2005, Kashlinksy, see (Reddy, 2016), "led a team of astronomers using NASA Spitzer telescope to explore the background glow of infrared light in one part of the sky. The researchers reported excessive patchiness in the glow and concluded it was caused by the aggregate light of the first sources to illuminate the universe more than 13 billion years ago. Follow up studies confirmed that this cosmic infrared background (CIB) showed similar unexpected structures in other parts of the sky."

In 2013 another study by NASA, elucidated that the irregular glow of low energy X-rays in the cosmic X-ray background (CXB) matched the patchiness of the cosmic infrared background (CIB) quite well. The research team concluded, "the only object we know of that could be sufficiently luminous across this wide an energy range is a black hole", and that Primordial Black Holes must have been abundant among the earliest stars. They make up at least one out of every five of the sources contributing to the cosmic infrared background (CIB).

They go on to explain that the LIGO observation of the two Black Holes merging some 1.3 billion light years ago was the first ever detection of gravitational waves. This was the first direct detection of Black Holes. The masses of the Black Holes were 29 and 36 times the mass of our Sun. These masses are both unexpectedly large and surprisingly similar. This adds credence to Primordial Black Holes having properties quite similar to those detected by LIGO.

Again Juan Garcia-Bellido, see (Garcia-Bellido, 2017), in the "Journal of physics, 2017", states that "Massive Primordial Black Holes (MBPH) can be formed after inflation due to broad peaks in the primordial curvature power spectrum that collapse gravitationally during the radiation era, generating a broad mass-spectrum of Black Holes with masses ranging from 0.01 to 100,000 times our Sun's mass. If there are enough of these Massive Primordial Black Holes (MPBH), they can constitute the bulk of dark matter today. Such Primordial Black Holes could also be responsible for the observed fluctuations in the cosmic infrared background (CIB) and the cosmic X-ray background, as discussed earlier."

Countering this, on June 14th, 2019, Benjamin V. Lehmann and co-workers, see (Lehmann, 2019), submitted an article to "Journal of Cosmology and Astroparticle Physics (JCAP)" which states that "the process of Hawking radiation could halt at scales approaching the Planck length. In this case it would leave Black Holes that could constitute the entirety of dark matter, and that such a relic would not be detectable using known techniques."

In conclusion there are a significant number of scientists that believe that Primordial Black Holes could be the leading candidate for dark matter. To date, this theory remains unproven; however, there are many good reasons to pursue this approach, the most advantageous being that we would not have to introduce any new matter to explain dark matter.

CHAPTER 24

CONSEQUENCES OF PRIMORDIAL BLACK HOLES AS DARK MATTER

If it is confirmed in the future that most of dark matter is from Primordial Black Holes, then this leads to a surprising conclusion: that the Big Bang would also result in a Big Crunch. Let us start with a universe that has 72% dark energy and 28% dark matter (of which only ~5% is ordinary matter), for a total of 100%. It follows that 85% of all matter in the universe is dark matter, and ordinary matter only accounts for 15% of the remaining matter. If all dark matter is made up of Primordial Black Holes, then this gives the ratio of Primordial Black Holes (dark matter) to ordinary matter at an estimated 5.6 to 1.

Let V be the initial velocity of expansion of the universe, then as shown in Figure 1, the universe in the first second of expansion would contain 5.6 Primordial Black Holes to every 1 region going on to become ordinary matter.

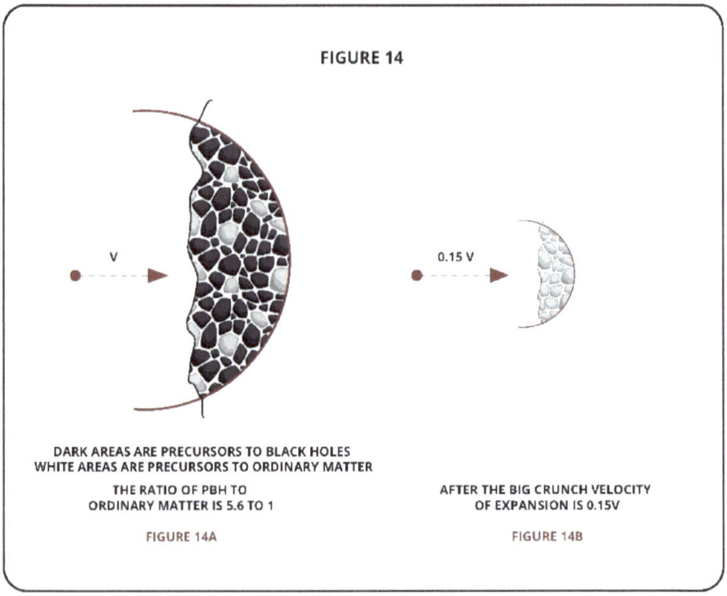

Figure 14: The initial expansion of the universe could contain 5.6 PBHs to every 1 region that forms ordinary matter. Figure 1B., shows the contraction of the universe due to PBHs.

If 85% of matter is made up of Primordial Black Holes, this would result in the velocity of expansion being reduced by 85%, as shown in Figure 14b. This figure shows that such a change in the velocity of expansion would cause a huge implosion of the universe and the energy of this contraction would be so big as to rival the Big Bang itself. This phenomenon is diametrically opposite to the Big Bang and is appropriately called the "Big Crunch".

If M is the total mass of the universe, then the kinetic energy, denoted by E, of the universe would be

$$E = \frac{1}{2}MV^2$$

where M is the sum of the mass of dark matter with the mass of ordinary matter, and V is the initial velocity of expansion, as explained in the introduction. Let m_1 represent the mass of ordinary matter and m_2 the mass of dark matter. Then,

$$E = \frac{1}{2}(m_1 + m_2)V^2$$

where $M = m_1 + m_2$ is the total mass.

However, because of the formation of all these Primordial Black Holes the velocity of the expansion would be reduced by 85%, and result in a velocity that is only 15% of the initial velocity. The new velocity of expansion would only be $0.15V$. This would cause a huge and sudden contraction of the universe and would be the Big Crunch. This is shown in Figure 14b.

The kinetic energy of the universe would now be the initial kinetic energy of the universe, less the kinetic energy of the contraction, or

$$E = \frac{1}{2}(m_1 + m_2)V^2 - \frac{1}{2}(m_1 + m_2)(0.85V)^2$$

The term, $\frac{1}{2}(m_1 + m_2)(0.85V)^2$, would represent the amount of energy that results from the abrupt contraction of the universe due to the formation of such a large amount of Primordial Black Holes. If we call the second term dark energy, this would represent a huge amount of energy (85% of the energy in the Big Bang). Then, because of this pullback of the expansion, this dark energy comes out to 72.2% of the energy in the universe. This is precisely the amount of dark energy that we have been searching for!

Since the Big Crunch moves a lot of the mass backwards, like a pendulum, it will move kinetic energy into potential energy, (see Figure 15). This potential energy (synonym for dark energy) would be stored in both dark matter and ordinary matter, but mainly in the Black Holes (85%) that would represent dark matter.

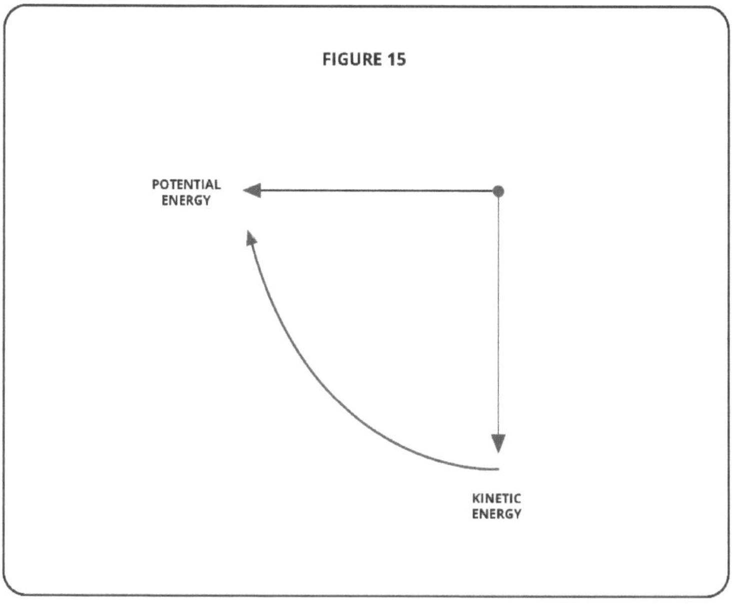

Figure 15: Since the Big Crunch moves a lot of the mass backwards, like a pendulum, it will move kinetic energy into potential energy.

In time this potential energy would be released back into the universe and be perceived as the universe speeding up. If the universe had more than 85% of dark matter, then the Big Crunch would be even bigger. Ultimately, if all matter was converted into dark matter (i.e., 100% of matter is Primordial Black Holes) the universe would not have formed at all. Figure 16 shows % dark energy vs % dark matter; So, at 100% dark matter there would be 100% dark energy and the universe would not be created at all. However, at 85% dark matter we get 68% dark energy. Figure 16 below shows the relationship of dark matter to dark energy.

FIGURE 16

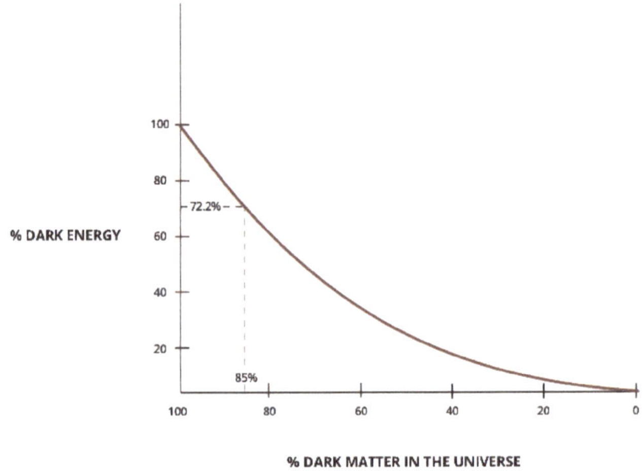

Figure 16: Shows the relationship of Dark Energy to Dark Matter in the form of Primordial Black Holes.

Thus, all the Dark Energy could be the Potential Energy formed by the "Big Crunch." In time, the release of this energy would be perceived that the universe is speeding up. Once all this energy is released back into the universe the final velocity would approach the initial velocity *V*, as shown in Figure 17 below.

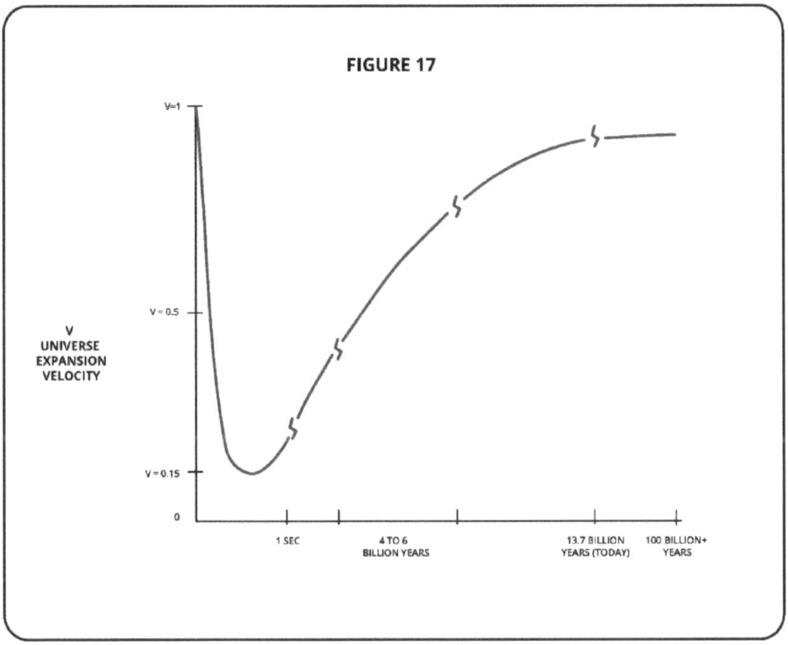

Figure 17: shows the velocity of expansion of the universe within the first second of the Big Bang ; the contraction in velocity due to the Big Crunch; and the subsequent increase in velocity over time.

Our equations indicate that there would be oscillations in the CMB that maybe detected by the James WEB probe. We do not know the extent of these oscillations, and this is something that would need to be worked out.

Chapter 25
Application to the Actual Universe

To apply this principle to the actual universe we must account for the fact that many black holes and supermassive black holes were formed after the Primordial black holes were formed. Supermassive Black Holes (SMBHs), as the name implies, contain between a million and a billion times more mass than a typical stellar Black Hole. They are thought to exist at the center of most large galaxies, including the center of our own galaxy, the Milky Way. Most recently, direct evidence for the existence of a Supermassive Black Holes (SMBHs) at the center of our galaxy has come from observations of orbiting stars. The high orbital velocities of these stars can only be explained by the presence of a Supermassive Black Hole (SMBHs).

Thus, the dark energy released by the Primordial Black Holes (PBHs) and other Black Holes created after the Big Crunch would depend on the ratio of Primordial Black Holes (PBHs) to the Black Holes created later in the universe. The total dark energy would be the same because the contraction of spacetime would be the same no matter when these Black Holes are created.

For example, suppose we say that Primordial Black Holes (PBHs) make up 75% of dark matter rather than 85%, as discussed earlier. Then, the dark energy formed from the Primordial Black Holes (PBHs) would drop from 68% to 56.2% (refer to Figure 16). The remaining dark energy, or 11.8%, would come from the formation of Supermassive Black Hole (SMBHs) and other Black Holes, formed at the later stages of the evolution of the universe.

It is generally assumed that most galaxy formation and filament galaxy formation would be mature after 4 to 6 billion years. Thus, most of the Supermassive Black Hole (SMBH) and other Black Hole formation would be completed in the first 4 to 6 billion years after the Big Bang. By this time, most galaxies would have a Supermassive Black Hole (SMBH) at their center. Although this process is still ongoing, we can assume for purposes of this discussion that it is 98% complete and this would not affect our conclusion.

Thus, with Primordial Black Holes (PBHs) at 75% of dark matter, the minimal velocity would be 0.25 of the initial velocity of expansion of the universe. As shown in Figure 18, within the first second the velocity would drop to 0.25V, then the expansion would continue, but at a rate slower than shown in Figure 17. This is because the expansion is being slowed by the continued formation of Supermassive Black Holes (SMBHs) and other Black Holes. After 4 to 6 billion years, if we assume that the bulk of Black Holes are formed, then the velocity of expansion of the universe would start to increase at a faster rate and the universe will be perceived to speed up. As shown in Figure 18, the velocity of expansion would just be catching up to the initial velocity of expansion, which would ultimately be achieved 100+ Billion years or later. The exact rate of increase and slope of this curve will depend upon how well the ratio of Primordial Black Holes (PBHs) to subsequent Black Hole formation are defined. Hence, our model of the Universe predicts at least two concentric rings, and maybe more, in the CMB, (see Illustration 27).

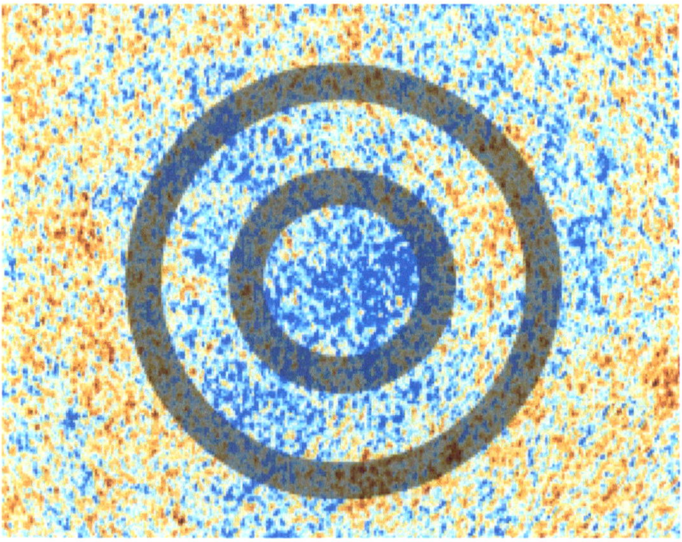

Illustration 27: This is an illustration of the concentric rings on the cosmic microwave background (CMB), that was first discovered by Gurzadyan and Penrose (2010). The role of these rings is reinterpreted here, and fits the big crunch scenario proposed. The model proposed here predicts at least two such rings.

It should be noted, many cosmologists agree with the fact that after 4 to 6 billion years the universe started speeding up. Thus, the curve shown in Figure 18, agrees quite well with their expectations.

FIGURE 18

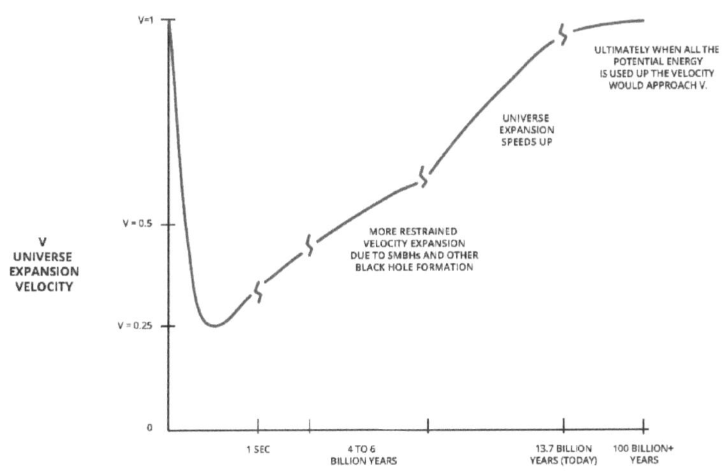

Figure 18: shows the velocity of the universe during the first second; the more restrained expansion of the universe due to SMBH and other black hole formation during the first 4 to 6 billion years; then the universe expansion speeding up after this period of time to the present time of 13.7 billion years; and eventually approaching the initial velocity of expansion.

The model we propose predicts oscillations of the universe in the first second of expansion of the universe. Figure 19 (below) shows the velocity of the universe in the first second; the more restrained expansion of the universe during the first 4-6 billion years; then, the universe expansion speeding up after this period of time to the present time of 13.7 billion years; and eventually approaching the initial velocity of expansion.

FIGURE 19

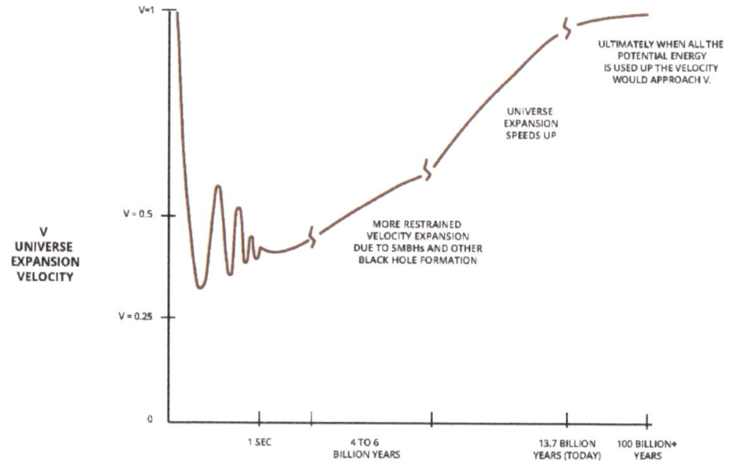

Figure 19: shows the velocity of the universe during the first second; the more restrained expansion of the universe due to SMBH and other black hole formation during the first 4 to 6 Billion years; then the universe expansion speeding up after this period of time to the present time of 13.7 Billion years; and eventually approaching the initial velocity of expansion. This could also be evidence of additional rings in the CMB that our model predicts. The model used here predicts oscillations of the universe.

The proposed model predicts that because of these oscillations, as shown in Fig. 19 above, we should be able to see oscillations in the universe in the first second of expansion. The model predicts at least two concentric rings in the CMB, and perhaps even more fainter rings (which may be seen by the new James Webb telescope) as it tries to map the CMB to less than 200 million years ago.

Although not proven, this approach, or other versions of this concept should be further investigated by scientists and cosmologists. It needs to be proven that Primordial Black Holes (PBHs) are most if not all dark matter. The ratio of Primordial Black Holes (PBHs) to subsequent Black Hole formation is a critical number and needs to be more precisely determined. Dynamic models of these concepts would need to be developed to simulate the Primordial Black Holes (PBH) contraction and then the subsequent formation of Supermassive Black Holes (SMBHs) and other Black Holes, after the Big Crunch.

Illustration 27 shows the concentric rings on the Cosmic Microwave Background (CMB), that was first discovered by Guarzadon & Penrose (2010).

CHAPTER 26
THE EFFECT OF ENCRUSTED DARK MATTER ON GALAXIES

Because of the Dark Matter encrusting the Milky Way, it will rotate our galaxy as if it were one mass, thus the arms of the galaxy will not curl up upon itself. Dark matter has been linked to galaxy rotation for decades (see, Figure 20). Astronomer Ver Rubin pioneered work on galaxy rotation rates, showing that spiral galaxies rotate faster than if their gravity were solely due to the constituent stars and gas. An additional invisible substance known as dark matter must influence galaxy rotation.

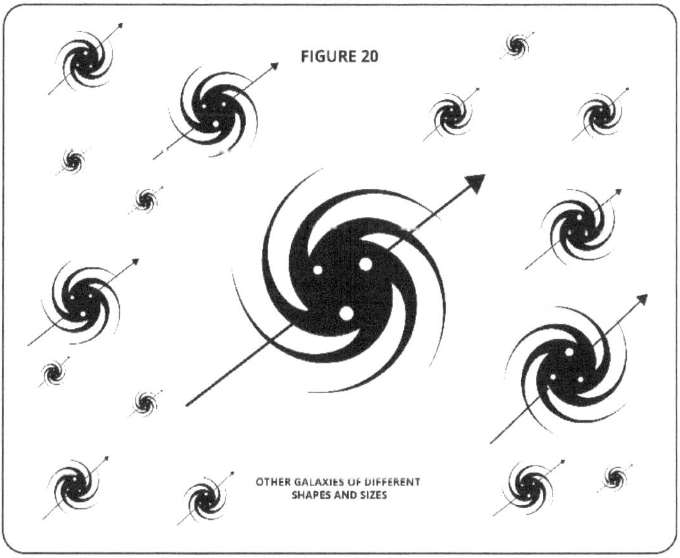

Figure 20: This is an illustration of galaxies spinning in opposite directions due to the randomness in the universe.

Furthermore, the proposed model would predict that because all galaxies are encrusted in dark matter, they will speed up the expansion of the universe, due to all the dark matter within them, as shown in Figure 21 below.

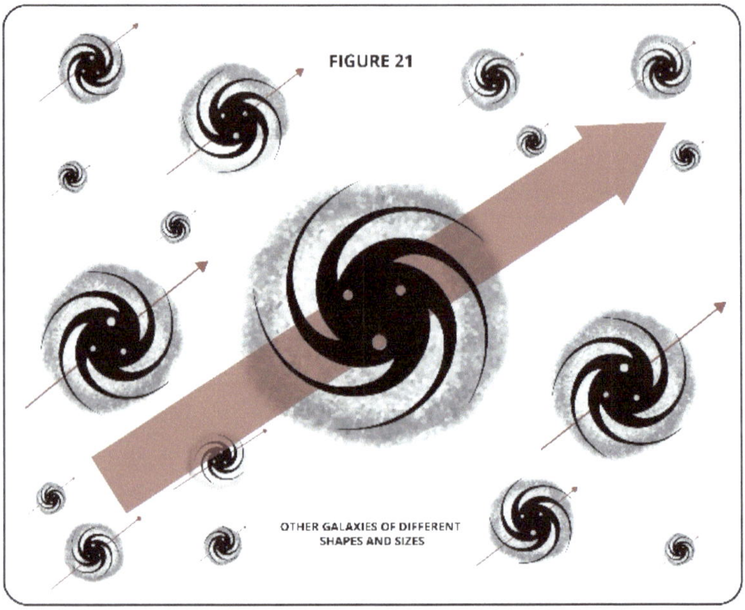

Figure 21: Illustration of galaxies, and how with the Dark Matter & Dark Energy in the Black Holes (due to the Big Crunch), they will speed up the expansion of the Universe.

The advantage of this approach is that we do not have to introduce any dark matter or dark energy to explain the Cosmos.

Conclusion

In conclusion, Cosmologists and scientists should increase their efforts to prove the existence of Primordial Black Holes (PBHs), as this could greatly simplify our model of the Universe. Having read the Book Proofs, the author feels that the result of this Model is so amazing-because it predicts the exact amount of Dark Matter (28%) and Dark Energy (72%) - that there is a high probability that the Model proposed in this book is correct!

References

Webb's Launch GSFC/NASA. (2022). https://www.jwst.nasa.gov/

Abbott, B. P. (2016). Observation of Gravitational Waves from a Binary Black Hole Merger. APS, 116(6). https://doi.org/10.1103/PhysRevLett.116.061102

Abraham Loeb. (2006). THE DARK AGES of the Universe. Scientific American, 295(No. 5). https://www.jstor.org/stable/26069038

Adam Becker. (2022). What Is Spacetime Really Made Of? - Scientific American. https://www.scientificamerican.com/article/what-is-spacetime-really-made-of/

Andrew J. Larkoski. (2019). Elementary Particle Physics: An Intuitive Introduction - Andrew J. Larkoski - Google Books. Cambridge University Press. https://books.google.com.au/books?id=aIOWDwAAQBAJ&printsec=frontcover&dq=fundamental+particle+physics&hl=en&sa=X&redir_esc=y

Bekenstein, J. D. (1973). Black holes and entropy. Physical Review D, 7(8), 2333–2346. https://doi.org/10.1103/PhysRevD.7.2333

Bertone, G., & Hooper, D. (2018). History of dark matter. Reviews of Modern Physics, 90(4), 045002. https://doi.org/10.1103/RevModPhys.90.045002

Bojowald, M. (2007). What happened before the Big Bang? Nature Physics, 3(8), 523–525. https://doi.org/10.1038/nphys654

Bowler, J. (2022). In a Wild Twist, Physicists Have Revived an Alternative

Theory of Gravity. https://www.sciencealert.com/physicists-have-just-revived-a-previously-debunked-alternative-theory-of-gravity

Boyer, T. (1985). The classical vacuum. JSTOR. https://www.jstor.org/stable/24967767

Dorminey, B. (2021). The Beginning to the End of the Universe: The mystery of dark energy | Astronomy.com. https://astronomy.com/magazine/news/2021/02/the-beginning-to-the-end-of-the-universe-the-mystery-of-dark-energy

Canetti, L., Drewes, M., & Shaposhnikov, M. (2012). Matter and antimatter in the universe. New Journal of Physics, 14(9), 095012. https://doi.org/10.1088/1367-2630/14/9/095012

Carlisle, C. M. (n.d.). Could Dark Matter Be Black Holes? - Sky & Telescope - Sky & Telescope. Retrieved June 19, 2022, from https://skyandtelescope.org/astronomy-news/black-holes/could-dark-matter-be-black-holes/

Carroll, B. W., & Ostlie, D. A. (2019). An Introduction to Modern Astrophysics, Second Edition. In An Introduction to Modern Astrophysics, Second Edition. https://doi.org/10.1017/9781108380980

Carroll, S. M. (2004). Why is the Universe Accelerating? AIP Conference Proceedings, 743(1), 16–32. https://doi.org/10.1063/1.1848314

Carroll, S. M. (2006). Is our Universe natural? In Nature (Vol. 440, Issue 7088, pp. 1132–1136). Nature Publishing Group. https://doi.org/10.1038/nature04804

Choi, C. Q., & Harvey, A. (2022). Our expanding universe: Age, history & other facts. Space. https://www.space.com/52-the-expanding-universe-from-the-big-bang-to-today.html

O'Luanaigh, C. (2013). Highlights from CERN in 2013 | CERN. https://home.cern/news/news/accelerators/highlights-cern-2013

O'Luanaigh, C. (2014). Fifty years of quarks | CERN. https://home.cern/news/news/physics/fifty-years-quarks

Clesse, S., & García-Bellido, J. (2018). Seven hints for primordial black hole dark matter. Physics of the

Dark Universe, 22, 137–146. https://doi.org/10.1016/j.dark.2018.08.004

Davies, P. C. W. (1999). The fifth miracle: the search for the origin and meaning of life. Simon & Schuster.

de Felice, A., & Tsujikawa, S. (2010). f (R) theories. In Living Reviews in Relativity (Vol. 13, Issue 1). Albert Einstein Institut. https://doi.org/10.12942/lrr-2010-3

Deabreu, A., Contreras, D., & Scott, D. (2015). Searching for concentric low variance circles in the cosmic microwave background. Journal of Cosmology and Astroparticle Physics, 2015(12). https://doi.org/10.1088/1475-7516/2015/12/031

Del Popolo, A. (2013). Non-baryonic dark matter in cosmology. AIP Conference Proceedings, 1548(1), 2–63. https://doi.org/10.1063/1.4817029

Overbye, D. (2015). The Flip Side of Optimism About Life on Other Planets - The New York Times. https://www.nytimes.com/2015/08/04/science/space/the-flip-side-of-optimism-about-life-on-other-planets.html

Eisenstaedt, J. (1989). The early interpretation of the Schwarzschild solution. https://philpapers.org/rec/EISTEI

Ellis, G. F. R., & Uzan, J.-P. (2005). c is the speed of light, isn't it? American Journal of Physics, 73(3), 240–247. https://doi.org/10.1119/1.1819929

Ferris, T. (1988). Coming of age in the Milky Way. In camw. Morrow.

Reddy, F. (2019). Possible Link Between Primordial Black Holes and Dark Matter | NASA. https://www.nasa.gov/feature/goddard/2016/nasa-scientist-suggests-possible-link-between-primordial-black-holes-and-dark-matter

Close, F. (2009). Antimatter. https://books.google.com.au/books?hl=en&lr=&id=lLRwDwAAQBAJ&oi=fnd&pg=PP1&dq=what+is+antimatter&ots=089lUsDekh&sig=M46wOlzV1kuFpfeOlhrtgQQG0zU

Mamoun, F. (2021). Black holes and dark matter — are they one and the same? | Department of Physics. https://physics.yale.edu/news/black-holes-and-dark-matter-are-they-one-and-same

G. M. McCracken, Garry McCracken, P. S. (2005). Fusion: The Energy of the Universe - G. M. McCracken, Garry McCracken, Peter Stott - Google Books. Academic Press. https://books.google.com.au/books?hl=en&lr=&id=wL6ECJexkqUC&oi=fnd&pg=PA1&dq=nuclear+fusion+early+universe&ots=VepG9dUSQF&sig=IMbUEXFuodEHdRrWzC4J8uihWnE

García-Bellido, J. (2017). Massive Primordial Black Holes as Dark Matter and their detection with Gravitational Waves. Journal of Physics: Conference Series, 840(1), 012032. https://doi.org/10.1088/1742-6596/840/1/012032

García-Bellido, J. (2019). Primordial black holes and the origin of the matter-antimatter asymmetry. Royalsocietypublishing.Org, 377(2161). https://doi.org/10.1098/rsta.2019.0091

García-Bellido, J., & Clesse, S. (2017). Black Holes from the Beginning of Time: A hidden population of black holes born less than one second after the big bang could solve the mystery of dark matter. Scientific American, 317(1), 38–43. https://doi.org/10.1038/scientificamerican0717-38

Gell-Mann, M. (2010). QUARKS. 212–240. https://doi.org/10.1142/9789812836854_0015

Lewis, G. (2016). Timeline: the history of gravity. https://theconversation.com/timeline-the-history-of-gravity-54528

Geroch, R. (1968). What is a singularity in general relativity? Annals of Physics, 48(3), 526–540. https://doi.org/10.1016/0003-4916(68)90144-9

Gibbs, P. (2012). The Universe- an Effect without Cause. Prespacetime Journal, 3(No 12).

Gribbin, J. (1993). Pay attention, Albert Einstein! New Scientist. https://www.newscientist.com/article/
mg13718543-900-pay-attention-albert-einstein/

Hawking, S., & Mlodinow, L. (2012). The grand design. Bantam Books.

Hossenfelder, S. (2013). Can we unify quantum mechanics and gravity? Physics World. https://physicsworld.com/a/can-we-unify-quantum-mechanics-and-gravity/

Howell, E., & May, A. (2022). What Is the Big Bang Theory? Space. https://www.space.com/25126-big-bang-theory.html

Hupp, E., Roy, S., & Watzke, M. (2006). NASA - NASA Finds Direct Proof of Dark Matter. NASA. https://www.nasa.gov/home/hqnews/2006/aug/HQ_06297_CHANDRA_Dark_Matter.html

Deaton, J. (2019). Einstein showed Newton was wrong about gravity. Now scientists are coming for Einstein. https://www.nbcnews.com/

mach/science/einstein-showed-newton-was-wrong-about-gravity-now-scientists-are-ncna1038671
Gregory, J. C. (1972). A Short History of Atomism: From Democritus to Bohr. A. & C. Black, Limited. https://books.google.com.au/books?id=uxpSAQAAMAAJ&q=history+of+the+hydrogen+atom&dq=history+of+the+hydrogen+atom&hl=en&sa=X&redir_esc=y
Kashlinsky, A. (2016). LIGO GRAVITATIONAL WAVE DETECTION, PRIMORDIAL BLACK HOLES, AND THE NEAR-IR COSMIC INFRARED BACKGROUND ANISOTROPIES. The Astrophysical Journal, 823(2), L25. https://doi.org/10.3847/2041-8205/823/2/l25
Croswell, K. (2020). Astronomers have found the edge of the Milky Way at last | Science News. https://www.sciencenews.org/article/astronomers-have-found-edge-milky-way-size
Kolb, E. W., & Turner, M. S. (2018). The early universe. CRC Press. https://doi.org/10.1201/9780429492860
Krauss, L. M. (2012). A Universe from Nothing: Why There Is Something Rather than Nothing - Lawrence M.
Krauss - Google Books. https://books.google.com.au/books?hl=en&lr=&id=nfGlsiDGbxkC&oi=fnd&pg=PA13&dq=where+did+the+matter+came+from+in+the+universe&ots=UT7tzEYa5Y&sig=QU2HluWQir8FRX22W274UzGKFJo
Lahav, O. (2020). Dark Energy: is it 'just' Einstein's Cosmological Constant ? Contemporary Physics, 61(2), 132–145. https://doi.org/10.1080/00107514.2020.1837456
Lehmann, B. V., Johnson, C., Profumo, S., & Schwemberger, T. (2019). Direct detection of primordial black hole relics as dark matter. Journal of Cosmology and Astroparticle Physics, 2019(10). https://doi.org/10.1088/1475-7516/2019/10/046
Williams, M. (2019). Now We Know That Dark Matter Isn't Primordial Black Holes - Universe Today. https://www.universetoday.com/141923/now-we-know-that-dark-matter-isnt-primordial-black-holes/
Fore, M. (2018). Is Dark Matter Made Up of Mini Black Holes from the Big Bang? | Live Science. https://www.livescience.com/62235-dark-matter-primordial-black-holes.html
Schirber, M. (2018). Physics - Controversy Continues over Black Holes

as Dark Matter. https://physics.aps.org/articles/v11/99

Moskvitch, K. (2018). Troubled Times for Alternatives to Einstein's Theory of Gravity. Quanta Magazine. https://www.quantamagazine.org/troubled-times-for-alternatives-to-einsteins-theory-of-gravity-20180430/

Murgia, R., Scelfo, G., Viel, M., & Raccanelli, A. (2019). Lyman- Forest Constraints on Primordial Black Holes as Dark Matter. Physical Review Letters, 123(7). https://doi.org/10.1103/PhysRevLett.123.071102

Musser, G. (2010). "Twistor" Theory Reignites the Latest Superstring Revolution. Scientific American. https://www.scientificamerican.com/article/simple-twist-of-fate/

O'Raifeartaigh, C., O'Keeffe, M., Nahm, W., & Mitton, S. (2017). Einstein's 1917 static model of the universe: a centennial review. European Physical Journal H, 42(3), 431–474. https://doi.org/10.1140/epjh/e2017-80002-5

Oba, Y., Takano, Y., Furukawa, Y., Koga, T., Glavin, D. P., Dworkin, J. P., & Naraoka, H. (2022). Identifying the wide diversity of extraterrestrial purine and pyrimidine nucleobases in carbonaceous meteorites. Nature Communications, 13(1). https://doi.org/10.1038/s41467-022-29612-x

Ouellette, J. (2015). In Memoriam: Jacob Bekenstein (1947–2015) and Black Hole Entropy. https://blogs.scientificamerican.com/cocktail-party-physics/in-memoriam-jacob-bekenstein-1947-2015-and-black-hole-entropy/

Overbye, D. (2003). Where Are Those Aliens? - The New York Times. https://www.nytimes.com/2003/11/11/science/where-are-those-aliens.html

Overbye, D. (2013). Far-Off Planets Like the Earth Dot the Galaxy - The New York Times. https://www.nytimes.com/2013/11/05/science/cosmic-census-finds-billions-of-planets-that-could-be-like-earth.html

Overbye, D. (2017). Cosmos Controversy: The Universe Is Expanding, but How Fast? - The New York Times. The New York Times. https://www.nytimes.com/2017/02/20/science/hubble-constant-universe-expanding-speed.html

P. James, E. Peebles, Lyman A. Page, Jr., R. B. P. (2009). Finding the Big

Bang - P. James E. Peebles, Lyman A. Page, Jr., R. Bruce Partridge - Google Books. https://books.google.com.au/books?hl=en&lr=&id=QdcdcMi-xpgC&oi=fnd&pg=PR11&dq=big+bang&ots=jWjv-_QOyA&sig=jcXPQJSfjtl0cI3zrB3Oz6KFp6w

Pallis, C., & Toumbas, N. (2017). Starobinsky Inflation: From Non-SUSY to SUGRA Realizations. Advances in High Energy Physics, 2017. https://doi.org/10.1155/2017/6759267

Penrose, R. (2016). Fashion, Faith, and Fantasy in the New Physics of the Universe. Princeton University Press. https://doi.org/10.1515/9781400880287

Penrose, R. (2015). Palatial twistor theory and the twistor googly problem. Royalsocietypublishing.Org, 373(2047). https://doi.org/10.1098/rsta.2014.0237

Penrose, R. (2016). The Emperor's New Mind: Concerning Computers, Minds, and The Laws of Physics (Oxford Landmark Science). Oxford Univ Pr. http://amazon.co.jp/o/ASIN/0198784929/

Penrose, R. (2011). Cycles of time : an extraordinary new view of the universe. Alfred A. Knopf.

Pillitteri, A. (2017). Fundamental forces and particles. In The Universe Untangled Modern physics for everyone. IOP Publishing. https://doi.org/10.1088/978-1-6817-4513-8ch2

Ribeiro, M., Henriques, T., Castro, L., Souto, A., Antunes, L., Costa-Santos, C., & Teixeira, A. (2021). The Entropy Universe. Entropy, 23(2), 222. https://doi.org/10.3390/e23020222

Lighthouse, R. (2014). The First Periodic Table for Elementary Particles - Richard Lighthouse - Google Books. https://books.google.com.au/books?id=sGYMBgAAQBAJ&pg=PA6&dq=leptons+(green),+quarks+(purple)+and+bosons&hl=en&sa=X&ved=2ahUKEwi6vvSctMf4AhXwSWwGHaYICswQ6wF6BAgAEAI

Ridley, M. (2006). Genome : the autobiography of a species in 23 chapters.

Sabulsky, D. O., Dutta, I., Hinds, E. A., Elder, B., Burrage, C., & Copeland, E. J. (2019). Experiment to Detect Dark Energy Forces Using Atom Interferometry. Physical Review Letters, 123(6). https://doi.org/10.1103/PhysRevLett.123.061102

Shipman, H. ~L. (1975). The implausible history of triple star models for

Cygnus X-1 Evidence for a black hole. Astrophysical Letters, 16, 9–12. https://adsabs.harvard.edu/pdf/1975ApL....16....9S

Siegel, E. (2017). Ask Ethan: What Was the Entropy Of The Universe At The Big Bang? https://www.forbes.com/sites/startswithabang/2017/04/15/ask-ethan-what-was-the-entropy-of-the-universe-at-the-big-bang/?sh=36572f667280

Smolin, L. (2019). Einstein's Unfinished Revolution: The Search for What Lies Beyond the Quantum. In Angewandte Chemie International Edition, 6(11), 951–952.

Weinberg, S. (1994). Life in the Universe. In Scientific American. Scientific American. https://www.jstor.org/stable/24942868

Steven, T. C. (2017). Cosmic inflation debate bleeds into popular science media. Physics Today. https://doi.org/10.1063/pt.6.3.20170605a

Stoner, B. R., & Glass, J. T. (2012). Nanoelectronics: Nothing is like a vacuum. In Nature Nanotechnology (Vol. 7, Issue 8, pp. 485–487). Nature Publishing Group. https://doi.org/10.1038/nnano.2012.130

Štorchová, H. (1994). Gesteland, R.F., Atkins, J.F. (ed.): The RNA world. Biologia Plantarum, 36(3), 358–358. https://doi.org/10.1007/bf02920931

Sutter, P. (2022). Loop quantum gravity: Does space-time come in tiny chunks? Space. https://www.space.com/loop-quantum-gravity-space-time-quantized

Sutter, P. (2021). What is a singularity? Live Science. https://www.livescience.com/what-is-singularity

Sutter, P. (2022). What happens at the center of a black hole? Space. https://www.space.com/what-happens-black-hole-center

Sutter, P. (2020). Dangerous "naked" black holes could be hiding in the universe | . Live Science. https://www.livescience.com/do-naked-singularities-exist.html

Thompson, T. A., Kochanek, C. S., Stanek, K. Z., Badenes, C., Post, R. S., Jayasinghe, T., Latham, D. W.,

Bieryla, A., Esquerdo, G. A., Berlind, P., Calkins, M. L., Tayar, J., Lindegren, L., Johnson, J. A., W-S

Holoien, T., Auchettl, K., & Covey, K. (2019). Discovery of a Candidate Black Hole–Giant Star Binary System in the Galactic Field. In arxiv.org. https://arxiv.org/abs/1806.02751

Tillman, N. T., Dobrijevic, D., & Biggs, B. (2022). Black holes: Everything you need to know. Space. https://www.space.com/15421-black-holes-facts-formation-discovery-sdcmp.html

Tsujikawa, S. (2003). Introductory review of cosmic inflation. http://arxiv.org/abs/hep-ph/0304257

Tully, C. G. (2011). Elementary Particle Physics in a Nutshell. Princeton University Press. https://books.google.com.au/books?id=vLy2YlkXZuEC&printsec=frontcover&dq=basic+particle+physics&hl=en&sa=X&redir_esc=y

Valcin, D., Bernal, J. L., Jimenez, R., Verde, L., & Wandelt, B. D. (2020). Inferring the age of the universe with globular clusters. Journal of Cosmology and Astroparticle Physics, 2020(12), 002. https://doi.org/10.1088/1475-7516/2020/12/002

W. N. Cottingham, D. A. Greenwood, D. A. G. (2001). An Introduction to Nuclear Physics - W. N. Cottingham, D. A. Greenwood, Derek A. Greenwood - Google Books. Cambridge University Press. https://books.google.com.au/books?id=0VIpJPn-qWoC&pg=PA2&dq=introduction+to+fermion+and+boson+statistics&hl=en&sa=X&ved=2ahUKEwiCiMOWocj4AhXRmuYKHR4pBy8Q6AF6BAgJEAE

Webb, S. (2015). If the universe is teeming with aliens... where is everybody? seventy-five solutions to the fermi paradox and the problem of extraterrestrial life. https://link.springer.com/978-3-319-13236-5

Wood, C., & Stein, V. (2022). What is string theory? Space. https://www.space.com/17594-string-theory.html

Penrose, R. (1969). Gravitational Collapse: the Role of General Relativity. NCimR, 1, 252.

Emeka-Okafor, V. (2021). What are stellar-mass black holes? | Astronomy.com. https://astronomy.com/news/2021/09/what-are-stellar-mass-black-holes

Fryer, C. L. (2003). Black-hole formation from stellar collapse. Classical and Quantum Gravity, 20(10), S73. https://doi.org/10.1088/0264-9381/20/10/309

Betz, E. (2020). What Would Happen if You Fell into a Black Hole? Astronomy. https://astronomy.com/news/2020/07/what-would-happen-if-you-fell-into-a-black-hole

Koberlein, B. (2021). Hawking Made a Prediction About Black Holes, and Physicists Just Confirmed it - Universe Today. https://www.universetoday.com/151713/hawking-made-a-prediction-about-black-holes-and-physicists-just-confirmed-it/

Bekenstein, J. D. (2019). Black Holes and Entropy. In Jacob Bekenstein (pp. 307–320). WORLD SCIENTIFIC. https://doi.org/10.1142/9789811203961_0023

de Lazaro, E. (2021). Astronomers Find Two Supermassive Black Holes in NGC 7727 |. Sci-News.Com. http://www.sci-news.com/astronomy/two-supermassive-black-holes-ngc-7727-10322.html

Bartusiak, M. (2016). Schwarzschild Solution: A Historical Perspective. APS, 2016, H5.001.

Johnston, H. (2019). Black hole is hairless, reveals analysis of gravitational waves. Physics World. https://physicsworld.com/a/black-hole-is-hairless-reveals-analysis-of-gravitational-waves/

GW190521. (n.d.). Retrieved June 26, 2022, from https://www.ligo.org/detections/GW190521.php

Pultarova, T. (2021). Black holes may have existed since the beginning of time (and could explain dark matter mystery). Space. https://www.space.com/primordial-black-holes-explain-dark-matter-universe-mysteries

Shears, T. (2012). The Standard Model. Philosophical Transactions of the Royal Society A: Mathematical, Physical and Engineering Sciences, 370(1961), 805–817. https://doi.org/10.1098/rsta.2011.0314

Siegel, E. (2019). What Came First: Inflation or The Big Bang? https://www.forbes.com/sites/startswithabang/2019/10/22/what-came-first-inflation-or-the-big-bang/?sh=312ecbf14153

Schmidt, B. P. (2012). Nobel lecture: Accelerating expansion of the Universe through observations of distant supernovae. Reviews of Modern Physics, 84(3), 1151–1163. https://doi.org/10.1103/RevModPhys.84.1151

Elizalde, E. (2021). The Hubble-Lemaître Law and the Expansion of the Universe. In The True Story of Modern Cosmology (pp. 137–166). Springer International Publishing. https://doi.org/10.1007/978-3-030-80654-5_5

M. Gomez, O. (2018). Five mysteries the Standard Model can't explain | symmetry magazine. https://www.symmetrymagazine.org/article/five-mysteries-the-standard-model-cant-explain

Pinfold, J. L. (2010). Dirac's dream - The search for the magnetic monopole. AIP Conference Proceedings, 1304(1), 234–239. https://doi.org/10.1063/1.3527206

Powell, C. S. (2015). Relativity versus quantum mechanics: the battle for the universe. The Guardian. https://www.theguardian.com/news/2015/nov/04/relativity-quantum-mechanics-universe-physicists

Penrose, R. (2012). The basic ideas of conformal cyclic cosmology. AIP Conference Proceedings, 1446(1), 233–243. https://doi.org/10.1063/1.4727997

Wood, C. (2019). What Is Quantum Gravity? Space. https://www.space.com/quantum-gravity.html

Durrer, R. (2015). The Cosmic Microwave Background: The history of its experimental investigation and its significance for cosmology. Classical and Quantum Gravity, 32(12). https://doi.org/10.1088/0264-9381/32/12/124007

Harwood, W. (2022). James Webb Space Telescope's optical alignment "perfect," NASA says -. CBS News. https://www.cbsnews.com/news/james-webb-space-telescope-nasa-optical-alignment/

Clegg, B. (2019). Dark Matter and Dark Energy: The Hidden 95% of the Universe. In Icon Books. https://books.google.com.au/books?id=hYCaDwAAQBAJ&printsec=frontcover&dq=review+of+dark+matter+and+energy&hl=en&sa=X&ved=2ahUKEwjun4Opo8r4AhVj4DgGHdDSC0QQ6AF6BAgCEAE

Betz, E. (2021). The Beginning to the End of the Universe: The Big Crunch vs. The Big Freeze. Astronomy.Com. https://astronomy.com/news/magazine/2021/01/the-beginning-to-the-end-of-the-universe-the-big-crunch-vs-the-big-freeze

Commissariat, T. (2016). LIGO detects first ever gravitational waves – from two merging black holes. Physics World. https://physicsworld.com/a/ligo-detects-first-ever-gravitational-waves-from-two-merging-black-holes/

WIKIPEDIA REFERENCES

(The materials in these references are subject to change but are constantly updated for accuracy and new knowledge.)

Big Bang. (n.d.). In *Wikipedia*. Retrieved January 17, 2022, from https://en.wikipedia.org/wiki?curid=6139891

Chronology of the universe. (n.d.). In *Wikipedia*. Retrieved January 17, 2022, from https://en.wikipedia.org/wiki?curid=19167840

Inflation (cosmology). (n.d.). In *Wikipedia*. Retrieved January 17, 2022, from https://en.wikipedia.org/wiki/Inflation_(cosmology)

Higgs boson. (n.d.). In *Wikipedia*. Retrieved January 17, 2022, from https://en.wikipedia.org/wiki/Higgs_boson

Primordial black hole. (n.d.). In *Wikipedia*. Retrieved January 17, 2022, from https://en.wikipedia.org/wiki/Primordial_black_hole

Fermion. (n.d.). In *Wikipedia*. Retrieved January 17, 2022, from https://en.wikipedia.org/wiki?curid=11529

Boson. (n.d.). In *Wikipedia*. Retrieved January 17, 2022, from https://en.wikipedia.org/wiki/Boson

Hadron. (n.d.). In *Wikipedia*. Retrieved January 17, 2022, from https://en.wikipedia.org/wiki/Hadron

Wilkinson Microwave Anisotropy Probe. (n.d.). In *Wikipedia*. Retrieved January 17, 2022, from https://en.wikipedia.org/wiki/Wilkinson_Microwave_Anisotropy_Probe

Thermodynamics. (n.d.). In *Wikipedia*. Retrieved January 20, 2022, from https://en.wikipedia.org/wiki/Thermodynamics

Entropy (classical thermodynamics). (n.d.). In *Wikipedia*. Retrieved January 20, 2022, from https://en.wikipedia.org/wiki/Entropy_(classical_thermodynamics)

Second law of thermodynamics. (n.d.). In *Wikipedia*. Retrieved January 20, 2022, from https://en.wikipedia.org/wiki/Second_law_of_thermodynamics

Entropy as an arrow of time. (n.d.). In *Wikipedia*. Retrieved January 20, 2022, from https://en.wikipedia.org/wiki/Entropy_as_an_arrow_of_time

Event horizon. (n.d.). In *Wikipedia*. Retrieved January 20, 2022, from https://en.wikipedia.org/wiki/Event_horizon

Cosmic microwave background. (n.d.). In *Wikipedia*. Retrieved January 20, 2022, from https://en.wikipedia.org/wiki/Cosmic_microwave_background

Thorne–Hawking–Preskill bet. (n.d.). In *Wikipedia*. Retrieved January 20, 2022, from https://en.wikipedia.org/wiki/Thorne%E2%80%93Hawking%E2%80%93Preskill_bet

f(R) gravity. (n.d.). In *Wikipedia*. Retrieved January 20, 2022, from https://en.wikipedia.org/wiki/F(R)_gravity

General relativity. (n.d.). In *Wikipedia*. Retrieved January 20, 2022, from https://en.wikipedia.org/wiki/General_theory_of_relativity

Newton's law of universal. gravitation. (n.d.). In *Wikipedia*. Retrieved January 20, 2022, from https://en.wikipedia.org/wiki/Newtonian_gravity

Gravitational singularity. (n.d.). In *Wikipedia*. Retrieved January 20, 2022, from https://en.wikipedia.org/wiki/Gravitational_singularity

Horizon problem. (n.d.). In *Wikipedia*. Retrieved January 20, 2022, from https://en.wikipedia.org/wiki?curid=695215

Inflation (cosmology). (n.d.). In *Wikipedia*. Retrieved January 20, 2022, from https://en.wikipedia.org/wiki/Inflation_(cosmology)

Black hole. (n.d.). In *Wikipedia*. Retrieved January 20, 2022, from https://en.wikipedia.org/wiki/Black_holes

Singularity (mathematics). (n.d.). In *Wikipedia*. Retrieved January 20, 2022, from https://en.wikipedia.org/wiki/Mathematical_singularity

Naked singularity. (n.d.). In *Wikipedia*. Retrieved January 20, 2022, from https://en.wikipedia.org/wiki/Naked_singularity

Tolman–Oppenheimer–Volkoff limit. (n.d.). In *Wikipedia*. Retrieved January 20, 2022, from https://en.wikipedia.org/wiki/ Tolman%E2%80%93Oppenheimer%E2%80%93Volkoff_limit

No-hair theorem. (n.d.). In *Wikipedia*. Retrieved January 20, 2022, from https://en.wikipedia.org/wiki/No-hair_theorem

Hawking radiation. (n.d.). In *Wikipedia*. Retrieved January 20, 2022, from https://en.wikipedia.org/wiki/Hawking_radiation

Lambda-CDM model. (n.d.). In *Wikipedia*. Retrieved January 20, 2022, from https://en.wikipedia.org/wiki/Lambda-CDM_model

Cold dark matter. (n.d.). In *Wikipedia*. Retrieved January 20, 2022, from

https://en.wikipedia.org/wiki/Cold_dark_matter
Expansion of the universe. (n.d.). In *Wikipedia*. Retrieved January 20, 2022, from https://en.wikipedia.org/wiki/Expansion_of_the_universe
Hubble's law. (n.d.). In *Wikipedia*. Retrieved January 20, 2022, from https://en.wikipedia.org/wiki/Hubble%27s_law
Cosmological constant. (n.d.). In *Wikipedia*. Retrieved January 20, 2022, from https://en.wikipedia.org/wiki/Cosmological_constant
Flatness problem. (n.d.). In *Wikipedia*. Retrieved January 20, 2022, from https://en.wikipedia.org/wiki/Flatness_problem
Magnetic monopole. (n.d.). In *Wikipedia*. Retrieved January 20, 2022, from https://en.wikipedia.org/wiki/Magnetic_monopole
Fine-tuned universe. (n.d.). In *Wikipedia*. Retrieved January 20, 2022, from https://en.wikipedia.org/wiki/Fine-tuned_universe
Planck constant. (n.d.). In *Wikipedia*. Retrieved January 21, 2022, from https://en.wikipedia.org/wiki/Planck_constant
Quantum. (n.d.). In *Wikipedia*. Retrieved January 21, 2022, from https://en.wikipedia.org/wiki/Quantum
Conformal cyclic cosmology. (n.d.). In *Wikipedia*. Retrieved January 21, 2022, from https://en.wikipedia.org/wiki/Conformal_cyclic_cosmology
Discovery of cosmic microwave background radiation. (n.d.). In *Wikipedia*. Retrieved January 21, 2022, from https://en.wikipedia.org/wiki/Discovery_of_cosmic_microwave_background_radiation
Black hole thermodynamics. (n.d.). In *Wikipedia*. Retrieved January 21, 2022, from https://en.wikipedia.org/wiki/Black_hole_thermodynamics
Dark matter. (n.d.). In *Wikipedia*. Retrieved January 21, 2022, from https://en.wikipedia.org/wiki/Dark_matter
Gravitational lens. (n.d.). In *Wikipedia*. Retrieved January 21, 2022, from https://en.wikipedia.org/wiki/Gravitational_lens
Baryonic dark matter. (n.d.). In *Wikipedia*. Retrieved January 21, 2022, from https://en.wikipedia.org/wiki/Baryonic_dark_matter
Primordial black hole. (n.d.). In *Wikipedia*. Retrieved January 21, 2022, from https://en.wikipedia.org/wiki/Primordial_black_hole
Particle physics. (n.d.). In *Wikipedia*. Retrieved March 7, 2022, from https://en.wikipedia.org/wiki/Particle_physics

Hydrogen atom. (n.d.). In *Wikipedia*. Retrieved March 7, 2022, from https://en.wikipedia.org/wiki/Hydrogen_atom

Particle decay. (n.d.). In *Wikipedia*. Retrieved April 10, 2022, from https://en.wikipedia.org/wiki/Particle_decay

Search for extraterrestrial intelligence. In *Wikipedia*. Retrieved May 22, 2022, from https://en.wikipedia.org/wiki/SETI

GLOSSARY

Aeon The period between successive and cyclic Big Bangs within the context of conformal cyclic cosmology.

Angular Momentum The rotational analog of linear momentum.

Anti-de Sitter Spacetime A spacetime with constant negative cosmological constant.

Anti-Higgs Field Hypothetical (Higgs) mechanism that removes some particles mass.

Anti-particle Has the same mass but with opposite physical charges (such as electric charge) of every type of particle in the universe.

Antimatter Matter composed purely of the anti-particles.

Baryonic Particles that constitute most of the mass of the observable matter in the universe and compose the nucleus of every atom.

Big Crunch A hypothetical scenario where gravitational attraction overcomes the expansion of the universe.

Billions The number 1 followed by 9 zeros, i.e., 1,000,000,000 or 10^9

Black Hole The result of a collapsed star so dense its gravitational pull is strong enough such that nothing can escape from it, not even light.

Bosons They are either force carrying, e.g., photons, or give rise to the phenomenon of mass, e.g., Higgs boson.

Celestial Objects Objects like moons, planets, stars, galaxies, nebulae, asteroids, planetary systems, star clusters, etc.

Conformal Cyclic Cosmology (CCC) A cyclic universe theory that consists of a infinite succession of aeons, each aeon starts with its own big bang and expands exponentially.

Cosmic Censorship Hypothesis The idea that no naked singularities exist in the universe.

Cosmic Infrared Background (CIB) Infrared radiation caused by stellar dust.

Cosmic Microwave Background (CMB) Faint cosmic background (electromagnetic) radiation from an early stage of the universe filling all space, from all directions.

Cosmic X-ray Background (CXB) X-ray emission produced primarily by emission from accreting Supermassive Black Holes.

Cosmological Constant The constant coefficient of a term added to the field equations of General Relativity.

Cosmological Inflation A famous theory of rapid exponential expansion of spacetime in the early universe.

Cosmological Principle The principle that the spatial distribution of matter in the universe is homogeneous and isotropic when considered on a large enough scale.

Dark Energy An, yet unknown form of energy that is thought to affect the universe on a large scale.

Dark Matter An unknown form of matter that is thought to account for approximately 85% of the matter in the universe.

Electromagnetic Field The combination of an electric field with a magnetic field, that not only propagates at the speed of light but can be identified as light.

Electron An elementary subatomic particle found in ordinary matter that has a function in numerous physical phenomena, such as electricity and magnetism.

Entropy A measure of the disorder or randomness in the universe (assumed a closed system), which is always increasing.

Epoch An era or age of the universe which is defined by a measurable change in nature.

European Organization for Nuclear Research (CERN) Organization that operates the largest particle physics collider in the world.

Event Horizon An invisible shell around a Black Hole which marks the point of no return for anything falling into it.

f(R) Gravity A modified theory of gravity which generalizes Einstein's general relativity in an attempt to solve open problems of the universe.

Fermions Fermion particles are made up of all quarks, leptons, and their odd number composites, such as all baryons and many atoms and nuclei.

Flatness Problem A cosmological fine-tuning problem associated the curvature of the geometry within the Big Bang model of the universe.

Four-dimensional Three spatial dimensions, in addition to time, making four-dimensions.

Galaxy A system of planets, suns, stars, asteroids, interstellar gas, dust, and dark matter, that is bound by gravity.

General Relativity Theory of gravity proposed by Albert Einstein in 1915.

General Theory of Relativity See, General Relativity.

Gravitational Constant An empirical physical constant of gravitational theory.

Gravitational Lensing Bending the light, due to gravitational effects, from the source as the light travels towards a distant observer.

Gravity A fundamental interaction resulting in mutual attraction between all objects with mass or energy.

Hadrons A composite subatomic particle consisting of two or more quarks, e.g., proton and the neutron.

Hawking Radiation A minuscule amount of thermal radiation emitted outside a Black Hole's event horizon.

Higgs Field A type of field in particle physics that plays a role in endowing some particles with mass.

Higgs Particle Named after Peter Higgs, it it the force carrying boson of the Higgs field.

Homogeneous The universe looks the same in all locations.

Horizon Problem The problem of apparent uniformity of matter and energy of the universe.

Hubble Constant The constant of proportionality associated with Hubble's law.

Hubble's Law Galaxies are observed to be moving away from Earth at speeds proportional to their distance, i.e., the farther they are, the faster they are moving away.

Isotropic The universe looks the same in all directions.

James Webb Space Telescope (JWST) The most powerful space telescope designed primarily to conduct infrared astronomy.

Kelvin The SI base unit of temperature. The Celsius equivalent is equal to the temperature in Kelvins minus 273.15.

Kerr Solution The Kerr metric is an exact solution of the Einstein field equations of general relativity, which describes the geometry of a rotating vacuum spacetime of a Black Hole.

Lambda-Cold Dark Matter (LCDM) The standard model of Big Bang cosmology in which the universe contains a cosmological constant (denoted by Lambda) associated with dark energy, cold dark matter (CDM), and ordinary matter.

Large Hadron Collider (LHC) The world's largest and highest-energy particle collider built by CERN (See, CERN).

Laser Interferometer Gravitational Observatory (LIGO) A large-scale astronomical tool designed to detect cosmic gravitational waves.

Leptons An elementary particle like an electron.

Light-year The distance that light travels in a vacuum in one Julian year (365.25 days)

Loop Quantum Gravity An attempt to unify quantum mechanics and general relativity to form a theory of quantum gravity.

Magnetic-monopole An, yet unobserved, elementary particle exhibiting one magnetic pole (north or south pole).

Massive Primordial Black Holes (MPBH) Black Holes of mass in the order of 10^6 solar masses or greater.

Matter Any substance that has mass and volume (occupying space).

Maxwell's Equations Equations of classical electromagnetism that describe how electric and magnetic fields are generated.

Milky Way The galaxy that includes our solar system.

Minkowski Spacetime A combination of three-dimensional flat space and time.

Modified Newtonian Theory (MOND) A modified version of Newton's law of universal gravitation accounting for observed properties of galaxies.

Naked Singularity A hypothetical gravitational singularity without an event horizon.

NASA The National Aeronautics and Space Administration responsible for space and other research.

NASA Spitzer Telescope It was an infrared space telescope launched in 2003 and ended on 30 January 2020.

NASA's Chandra X-ray Observatory an X-ray telescope launched in on July 23, 1999.

Newtonian Gravity A law of universal gravitation first published on 5 July 1687.

Non-baryonic Dark matter candidates such as axions, sterile neutrinos, weakly interacting massive particles, supersymmetric particles, geons, primordial Black Holes, etc.

Particle Physics A branch of physics that studies the nature of the particles that make up matter and radiation of the universe.

Photons A massless elementary particle that is the force carrier for the electromagnetic force, including electromagnetic radiation such as light and radio waves.

Planck Length Lengths of around 10^(–35) meters.

Planck Scale Quantities of space, time, energy and other units that are similar in magnitude to corresponding Planck units.

Planck's Constant A fundamental physical constant denoted by h.

Planet A suitably large astronomical body that is not a star nor artifact.

Primordial Black Hole (PBH) A conjectured type of black hole that is formed soon after the Big Bang.

Quantum Gravity A theory of gravity that adheres to the principles of quantum mechanics.

Quantum Mechanics A physical theory of nature at the scales of atoms and subatomic particles.

Quarks An elementary particle and a fundamental constituent of matter that does not exist in isolation.

Schwarzschild Radius The radius defining the event horizon of a Schwarzschild Black Hole.

Schwarzschild Solution An exact solution to the Einstein field equations that describes the gravitational field outside a (non-rotating) spherical mass.

Second Law of Thermodynamics The entropy of the universe tends to a maximum.

SI Units The International System of Units is the only official system of measurement.

Singularity A condition in which gravity increases so much that spacetime itself breaks down.

Solar Mass A standard unit of mass in astronomy approximately equal to 2×10^30 kilograms (approximately equal to the mass of the Sun).

Spacetime Theory that unifies the three dimensions of space and one dimension of time into a single four-dimensional spacetime.

Speed of light The speed of light in vacuum, which is a physical constant, is approximately 300000 kilometres per second or 186000 miles per second.

Spherical Symmetry The shape of a perfectly round object, resembling a ball.

Standard Model (LCDM) The Big Bang cosmological model of the universe which is consider the current standard model. See, (Lambda-Cold Dark Matter (LCDM)).

Star A celestial luminous spherical object of plasma held together by its own gravity, like our Sun.

Steady-state Universe A cosmological model where the density of matter in the expanding universe remains unchanged and looks the same at any time and any place, due to a continuous creation of matter.

Stellar Masses The mass of a star in units of solar mass. See (Solar mass).

String Theory A theory of quantum gravity where one-dimensional objects called strings replace point-like particles.

Sun The sun in our solar system.

Supermassive Black Hole Black Holes with mass in the order of millions to billions solar masses.

Thermal Equilibrium A system is said to be in thermal equilibrium with itself if the temperature of the system is uniform and constant.

Thermodynamics A branch of physics that deals with heat, work, and temperature, and their relation to the physical properties of matter and radiation.

Three-dimensional The three dimensions of space.

Tolman–Oppenheimer–Volkoff Limit The limit of the mass that a star reaches before it will collapse to a denser form.

Trillions The number 1 followed by 12 zeros, i.e., 1,000,000,000,000 or 10^12.

Twistor Theory A theory proposed by Roger Penrose to for quantum gravity.

Two-dimensional Surface Any surface like a tabletop, or the surface of a basketball.

Vacuum A region space devoid of matter.

Wilkinson Microwave Anisotropy Probe (WMAP) A NASA spacecraft that measured temperature differences across the sky in the cosmic microwave background (CMB) from 2001 to 2010.

Epilogue

The James Webb telescope has sent the recent pictures of galaxies 4 billion years ago with the highest resolution ever! These are the deepest view of the Universe ever captured and is the largest telescope ever built. It has 18 mirrors which had to be focused over 4 days on a single star for the telescope to be fully operational. Since then, the telescope has composed images (taken at different wavelengths over 12 hours) and takes us close to the edge of the Universe. It also showed water vapor in a distant star which is essential for life to exist. Another amazing fact is that we now think the Universe has maybe as much as 10 times more galaxies than previously thought!

www.ingramcontent.com/pod-product-compliance
Lightning Source LLC
LaVergne TN
LVHW051128210926
848482LV00001B/17
* 9 7 9 8 8 8 7 2 9 1 3 9 0 *